NUMBER POWER

A REAL WORLD APPROACH TO MATH

Financial Literacy

JERRY HOWETT

McGraw Hill Education

Bothell, WA • Chicago, IL • Columbus, OH • New York, NY

www.mheonline.com

 **Education**

Copyright © 2011, 2004 by The McGraw-Hill Companies, Inc.

Image Credits:

Cover 9 Comstock Images/Punchstock; **12** Nick Koudis/ Photodisc/Getty Images; **13** Ryan McVay/Photodisc/Getty Images; **22** Adam Crowley/Photodisc/Getty Images; **23** Andy Sotiriou/Photodisc/Getty Images; **25** C Squared Studios/ Photodisc/Getty Images; **32** Image100/SuperStock; **33** Ryan McVay/Digital Vision/Getty Images; **43** (bl)Ryan McVay/ Photodisc/Getty Images; **51** Allen Simon/Digital Vision/ Getty Images; **58** David De Lossy/Photodisc/Getty Images; **59** (bl)Image Source/Alamy; **70** Kim Steele/Blend Images/ Getty Images; **71** (bl)Janis Christie/Photodisc/Getty Images; **80** Steven Puetzer/Photographer's Choice/Getty Images; **81** Jack Star/PhotoLink/Getty Images; **91** Comstock Images/ Punchstock; **95** (br)Steve Cole/Photodisc/Getty Images; **96** Photodisc/Getty Images; **102** Randy Faris/CORBIS; **103** Keith Brofsky/Photodisc/Getty Images; **112** (bl)Comstock Images/Alamy; **113** Jeffrey Coolidge/Photodisc/Getty Images; **120** Medioimages/Photodisc/Getty Images; **121** (bl)Steve Cole/Photodisc/Getty Images; **130** Phillip Spears/Digital Vision/Getty Images; **131** Keith Brofsky/Photodisc/Getty Images; **137** Comstock Images/Punchstock; **140** (tr)Polka Dot Images/Jupiterimages, (bl)Stockbyte/Getty Images; **141** (bl) Comstock Images/Jupiterimages; **154** Digital Vision/Getty Images; **155** Comstock Images/Jupiterimages; **164** Diane Macdonald/Stockbyte/Getty Images; **165** Creatas/Punchstock; **174** Diane Macdonald/Photodisc/Getty Images; **175** Mikael Karlsson/Alamy; **188** UpperCut Images/SuperStock; **189** Stockbyte/Getty Images; **198** Jeffrey Hamilton/Digital Vision/Getty Images; **199** C. Sherburne/PhotoLink/Getty Images; **211** Comstock Images/Punchstock; **212** Stockbyte/ Getty Images; **223** Comstock Images/Punchstock; All artwork © The McGraw-Hill Companies, Inc.

Send all inquiries to:
Contemporary/McGraw-Hill
130 E. Randolph St., Suite 400
Chicago, IL 60601

ISBN: 978-0-07-661348-9
MHID: 0-07-661348-8

Printed in the United States of America.

1 2 3 4 5 6 7 8 9 10 RHR 15 14 13 12 11

TABLE OF CONTENTS

ABOUT THIS BOOK

This book has two main goals. The first goal is to illustrate and offer practice in the mathematics involved in being financially literate in the twenty-first century. The second goal is to introduce English language learners to the vocabulary that they need to work through each lesson.

The Pretest helps the student determine his or her skill level in the scope of financial literacy. An evaluation chart shows where each math skill is taught throughout the book.

Preceding each lesson, an exercise introduces the vocabulary used in the text. After each lesson is a vocabulary review and a language builder page that reviews the grammatical constructions used in the main body of the lesson. These three pages are designed to target the needs of ELL students and to support the transfer of mathematical skills from their native language to English.

The book has five main units based on CASAS competencies.

- **Managing Money** includes such everyday finance-related activities as earning, banking, eating, shopping, and saving.

- **Borrowing Money** includes details about the costs of taking out a loan and using a credit card.

- **Unavoidable Expenses** looks at the details of utility bills, insurance policies, and a variety of taxes.

- **Budgeting** reviews many of the topics from earlier lessons.

- **Math Skills Review** provides a review of basic math topics.

Before each math exercise there are references to the pages of the Math Skills Review section where the reader can review the mathematical skills that are needed to solve the problems in that exercise. This section also includes tips on using a calculator.

The Posttests give the reader a chance to measure the skills he or she has mastered and to decide whether some lessons need review. Finally, a Glossary lists the finance-related vocabulary that is used throughout the book.

After working through this book, every reader will be able to make better-informed decisions as a financially literate consumer.

Pretest

This is a test of the mathematical skills used in this book. Do every problem that you can. Answers are listed at the back of the book. The chart on page 8 lists the lessons where these skills are taught. You may use a calculator to solve these problems.

1. Round 128,703 to the nearest thousand.

2. What is $43,982 rounded to the nearest hundred dollars?

3. Steve drove about 12,000 miles last year and bought 436 gallons of gasoline. Find the fuel efficiency of his car to the nearest mile per gallon.

4. Mary and Tom pay $643 a month for their mortgage. How much do they pay for their mortgage in one year?

5. Jake sold three used cars in a month at the following prices: $4,599.00, $6,399.00, and $5,469.00. What was the average price of the cars?

6. In January Charlene paid $106 for her telephone bill. In February she paid only $62. What was her average telephone bill for those months?

7. Round 2.837 to the nearest tenth.

8. What is $39.568 rounded to the nearest cent?

9. Find the sum of 0.85, 1.3, and 9.418.

10. What is the difference between 0.76 and 0.3?

11. Find the product of $9.89 and 1.07 to the nearest cent.

12. Find the quotient of $21 \div 3.5$.

13. What is the cost of 2.15 pounds of cheese priced at $7.99 a pound?

14. At the beginning of March, Michelle had a balance of $887.43 in her checking account. During the month she wrote checks for $435, $58.27, and $127.96. She also deposited a check for $301.68. What was the balance in her account at the end of March?

15. Find the cost of 200 shares of stock that sell for 11.3 per share.

16. Molly's house has an assessed value of $115,000. The school tax rate in her community is 17 mills per dollar of assessed value. What is her yearly school tax?

17. Reduce $\frac{36}{54}$ to lowest terms.

18. What is $\frac{125}{275}$ reduced to lowest terms?

19. How many $\frac{1}{5}$-mile segments are there in 3 miles?

20. Adrienne's take-home pay is $1,580 a month. She spends an average of $316 a month on food for herself and her son. What fraction of her income does she spend on food?

21. Find 6.5% of $1,400.

22. What is 108% of $72?

23. 85 is what percent of 130? Find the answer to the nearest tenth of a percent.

24. $4,800 is what percent of $16,000?

25. Dinner for Bill and his wife cost $38.70. If Bill includes a 15% tip, what will be the total cost of the dinner?

26. Gloria makes $18.50 an hour. Her boss has given her a raise of 6%. What is her new hourly wage?

27. A jacket listed for $129 is on sale for 20% off. What is the sale price of the jacket?

28. Phil received a check for $120 in interest for a $2,000 bond. What rate of interest did he earn on the bond?

29. Pam paid $18,000 for her car. Five years later the car had a value of $6,800. By what percent did the value of the car depreciate? Round your answer to the nearest percent.

30. Rita and John bought their apartment for $65,000. They sold it five years later for $90,000. By what percent did the value of the apartment increase? Round your answer to the nearest percent.

31. Find the interest on $2,800 at 9% annual interest for three months.

32. What is the interest on $4,000 at 18% annual interest for one month?

33. Change 145 yards to feet.

34. Shirley's bus left at 10:40 in the morning and arrived at her destination at 2:12 in the afternoon. For how many hours was Shirley on the bus?

35. Change 2.5 pounds to ounces.

36. A half-gallon of milk costs $1.89. What is the cost per quart of the milk?

37. What is the perimeter of a basement floor that is 40 feet long and 24 feet wide?

38. What is the area of the floor in problem 37?

39. Find the area in square yards of the floor of a room that is 18 feet long and 12 feet wide.

40. Find the area of a triangle with a base of 30 feet and a height of 14 feet.

PRETEST EVALUATION CHART

Circle the number of any problem you missed. The column after the problem number tells you the lesson number where the skill is taught. The next column tells you the pages of the lesson.

Study the lessons that correspond to any of the problems that you got wrong.

Problem Number	Lesson Number	Lesson Pages	Problem Number	Lesson Number	Lesson Pages
1	1	11-20	21	11	111-118
2	1	11-20	22	11	111-118
3	2	21-30	23	12	119-128
4	2	21-30	24	12	119-128
5	3	31-40	25	13	129-138
6	3	31-40	26	13	129-138
7	4	41-48	27	14	139-152
8	4	41-48	28	14	139-152
9	5	49-56	29	15	153-162
10	5	49-56	30	15	153-162
11	6	57-68	31	16	163-172
12	6	57-68	32	16	163-172
13	7	69-78	33	17	173-186
14	7	69-78	34	17	173-186
15	8	79-90	35	18	187-196
16	8	79-90	36	18	187-196
17	9	93-100	37	19	197-209
18	9	93-100	38	19	197-209
19	10	101-110	39	20	212-221
20	10	101-110	40	20	212-221

UNIT 1
MANAGING
MONEY

The first eight lessons in this book cover the kind of financial situations consumers face nearly every day. The lessons include details about earning, banking, eating, shopping, repairing a home, and trying to save money. Pay close attention to the boldfaced vocabulary in each lesson to learn the terms and phrases commonly used and to achieve financial literacy.

LESSON 1: INCOME

Pre-Lesson Vocabulary Practice

Read the words and their meanings on the right. Next find them in the beginning of the lesson. Then carefully read the lesson material.

Use the boldfaced words on the right to complete the sentences below.

1. John makes a _____ of $18.55 an hour as a cook.

2. The amount a worker earns before any _____ are subtracted from his income is called gross pay.

3. Nathaniel works as a technician in a research lab. His _____ is $44,720.

4. Employers _____ certain amounts of money from employees' paychecks.

5. The _____ rate is often $1\frac{1}{2}$ times a worker's regular hourly wage.

6. Patricia works as a legal assistant. Her gross weekly _____ is $788.40.

Below is a list of words that appear throughout the lesson. Read each word and its definition. Then work with a partner to use the words in a sentence.

Social Security deductions – money subtracted from income and paid out when one retires

Medicare payments – income deductions used to pay for medical care at the age of 65 and beyond

federal income tax – an income deduction, money paid to the federal government for its operation

state income tax – a possible income deduction, money paid to the state government for its operation

$1\frac{1}{2}$ times – one and a half times

union – an organization that supports workers and helps to protect their rights

paycheck – the paper a bank uses to show payment of a worker's wage or salary

rate – a specific amount

% of – percent of or a part of

by law – according to what is legal

Internal Revenue Service (IRS) – the branch of the federal government that collects taxes

guidelines – help, information

income – an amount of money a person receives or earns, usually from working at a job

wage – income per hour, day, or year

yearly salary – income per year

overtime – extra, more than the regular amount of time worked

withhold – to take from, to subtract

deductions – amounts taken or subtracted from another amount

Income

Most people who get paid for their work earn an hourly **wage** or a **yearly salary.** The amount a worker earns before any **deductions** are subtracted from his **income** is called gross pay. The typical workweek in the United States is five 8-hour days, or 40 hours.

EXAMPLE 1 John makes a wage of $18.55 an hour as a cook. What is his gross weekly income for a 40-hour week?

Solution Multiply John's hourly wage by the number of hours he works each week.

$18.55 × 40 = **$742.00**

EXAMPLE 2 Sara is a high school Spanish teacher. Her yearly salary is $36,575. Find her weekly gross salary.

Solution Assume that a year has 52 weeks. Divide Sara's salary by 52.

$36,575 ÷ 52 = $703.365. . . or **$703.37**

Wage earners are usually paid a higher hourly rate if they work more than the normal 40-hour week. The **overtime** rate is often $1\frac{1}{2}$ times a worker's regular hourly wage.

EXAMPLE 3 John, the cook in Example 1, earns one and one-half times his regular wage when he works more than 40 hours in one week. How much did John make in overtime wages in a week when he worked 46 hours?

Solution John worked 46 − 40 = 6 hours of overtime. Multiply to find John's overtime wages.

6 × 1.5 × $18.55 = **$166.95**

Employers subtract or **withhold** certain amounts of money from an employee's paycheck. These **deductions** may include Social Security, Medicare payments, federal income tax, state income tax, contributions to retirement programs, union dues, and charitable contributions. The tables on page 15 are used to calculate how much money is withheld.

According to the Federal Insurance Contributions Act, or **F.I.C.A.,** 12.4% of every employee's income, up to an annual limit, must be contributed to **Social Security,** and 2.9% of the income must be contributed to **Medicare.** The total is 12.4% + 2.9% = 15.3% of an employee's income. By law, the employer pays half of this amount, and the employee pays the other half: 15.3% ÷ 2 = 7.65%.

EXAMPLE 4 For John, the cook in the first example, how much is withheld from his weekly check for Social Security and Medicare (F.I.C.A.) if he works 40 hours?

Solution Find 7.65% of $742.

0.0765 × $742 = $56.763 or **$56.76**

EXAMPLE 5 The state tax where Ricardo ate lunch is 4.5%. What was the tax on his lunch if his bill was $13.15?

Solution Find 4.5% of $13.15.

0.045 × $13.15 = $0.59175 or **$0.59**

EXAMPLE 6 How much income tax is withheld each week from a married person whose wages are $836 if the worker claims five withholding allowances?

Solution Look at the second table on page 15. $836 is between $830 and $840. The amount of income tax that is withheld with five withholding allowances is **$44.**

The amount a worker makes after deductions are subtracted from his gross pay is called his **net pay.** The net pay is the amount a worker takes home.

EXAMPLE 7 John, the cook in Example 1, is married, and he claimed three withholding allowances on his W-4 form. Find John's net weekly income for a 40-hour week if his employer deducts F.I.C.A., federal income tax, and $25 for John's retirement account.

Solution F.I.C.A. = 0.0765 × $742 = $56.763 or $56.76

$742 is between $740 and $750. With three withholding allowances, the income tax withheld is $48.

Total deductions = $56.76 + $48 + $25 = $129.76
Net pay = $742.00 − $129.76 = **$612.24**

Earnings	rate	hours	this period
Regular	$18.55	40.00	742.00
Overtime	---	---	---
Gross Pay			**$742.00**

Deductions		
F.I.C.A.		−56.76
Income Tax		−48.00
Retirement Contribution		−25.00
Net Pay		**$612.24**

Some people who sell things for a living receive a **commission** rather than a wage or a salary. A commission is a percent of total sales.

EXAMPLE 8 Charlie sells cars at a commission rate of 4.5%. What was his gross monthly income if he sold cars for a total value of $35,603.90 this month?

Solution Find 4.5% of $35,603.90.

4.5% = 0.045

0.045 × $35,603.90 = $1,602.1755 or **$1,602.18**

SINGLE Persons—WEEKLY Payroll Period
(For Wages Paid Through December 2004)

If the wages are—		And the number of withholding allowances claimed is—										
At least	But less than	0	1	2	3	4	5	6	7	8	9	10
		The amount of income tax to be withheld is—										
$600	$610	$78	$67	$58	$50	$41	$32	$23	$14	$8	$2	$0
610	620	80	69	60	51	42	33	24	15	9	3	0
620	630	83	70	61	53	44	35	26	17	10	4	0
630	640	85	72	63	54	45	36	27	18	11	5	0
640	650	88	73	64	56	47	38	29	20	12	6	0
650	660	90	75	66	57	48	39	30	21	13	7	1
660	670	93	78	67	59	50	41	32	23	14	8	2
670	680	95	80	69	60	51	42	33	24	15	9	3
680	690	98	83	70	62	53	44	35	26	17	10	4
690	700	100	85	72	63	54	45	36	27	18	11	5
700	710	103	88	73	65	56	47	38	29	20	12	6
710	720	105	90	75	66	57	48	39	30	21	13	7
720	730	108	93	78	68	59	50	41	32	23	14	8
730	740	110	95	80	69	60	51	42	33	24	15	9
740	750	113	98	83	71	62	53	44	35	26	17	10
750	760	115	100	85	72	63	54	45	36	27	18	11
760	770	118	103	88	74	65	56	47	38	29	20	12
770	780	120	105	90	75	66	57	48	39	30	21	13
780	790	123	108	93	78	68	59	50	41	32	23	14
790	800	125	110	95	80	69	60	51	42	33	24	15
800	810	128	113	98	83	71	62	53	44	35	26	17
810	820	130	115	100	85	72	63	54	45	36	27	18
820	830	133	118	103	88	74	65	56	47	38	29	20
830	840	135	120	105	90	75	66	57	48	39	30	21
840	850	138	123	108	93	78	68	59	50	41	32	23
850	860	140	125	110	95	80	69	60	51	42	33	24
860	870	143	128	113	98	83	71	62	53	44	35	26
870	880	145	130	115	100	85	72	63	54	45	36	27
880	890	148	133	118	103	88	74	65	56	47	38	29
890	900	150	135	120	105	90	76	66	57	48	39	30

MARRIED Persons—WEEKLY Payroll Period
(For Wages Paid Through December 2004)

If the wages are—		And the number of withholding allowances claimed is—										
At least	But less than	0	1	2	3	4	5	6	7	8	9	10
		The amount of income tax to be withheld is—										
$740	$750	$75	$66	$57	$48	$39	$30	$23	$17	$11	$5	$0
750	760	76	67	59	50	41	32	24	18	12	6	1
760	770	78	69	60	51	42	33	25	19	13	7	2
770	780	79	70	62	53	44	35	26	20	14	8	3
780	790	81	72	63	54	45	36	27	21	15	9	4
790	800	82	73	65	56	47	38	29	22	16	10	5
800	810	84	75	66	57	48	39	30	23	17	11	6
810	820	85	76	68	59	50	41	32	24	18	12	7
820	830	87	78	69	60	51	42	33	25	19	13	8
830	840	88	79	71	62	53	44	35	26	20	14	9
840	850	90	81	72	63	54	45	36	27	21	15	10
850	860	91	82	74	65	56	47	38	29	22	16	11
860	870	93	84	75	66	57	48	39	30	23	17	12
870	880	94	85	77	68	59	50	41	32	24	18	13
880	890	96	87	78	69	60	51	42	33	25	19	14
890	900	97	88	80	71	62	53	44	35	26	20	15
900	910	99	90	81	72	63	54	45	36	27	21	16
910	920	100	91	83	74	65	56	47	38	29	22	17
920	930	102	93	84	75	66	57	48	39	30	23	18
930	940	103	94	86	77	68	59	50	41	32	24	19
940	950	105	96	87	78	69	60	51	42	33	25	20
950	960	106	97	89	80	71	62	53	44	35	26	21
960	970	108	99	90	81	72	63	54	45	36	27	22
970	980	109	100	92	83	74	65	56	47	38	29	23
980	990	111	102	93	84	75	66	57	48	39	30	24
990	1,000	112	103	95	86	77	68	59	50	41	32	25
1,000	1,010	114	105	96	87	78	69	60	51	42	33	26
1,010	1,020	115	106	98	89	80	71	62	53	44	35	27
1,020	1,030	117	108	99	90	81	72	63	54	45	36	28
1,030	1,040	118	109	101	92	83	74	65	56	47	38	29

To solve the problems in the next exercise, review:

- multiplying decimals, page 231
- finding a percent of a number, page 236

EXERCISE 1

Part A

Use a calculator to solve the problems in this exercise. Use the two income tax withholding tables on page 15 when you need them. Remember that an employee's F.I.C.A. contribution is 7.65% of gross income.

1. Simon works as a security guard and makes $12.80 an hour. What is his gross weekly income for a 40-hour week?

2. Elena is a cashier in a convenience store. She makes $8.25 an hour for the first 40 hours and one and a half times her regular wage for overtime work. Find her gross weekly salary if she works 47 hours.

3. Nathaniel works as a technician in a research lab. His yearly salary is $44,720. What is his gross weekly salary?

4. Charlene works in a store that sells books and CDs. She gets a commission of 5% on all sales. What is her commission in a month when she sold books and CDs for a total of $21,518?

5. Myrna works as a housekeeper in a large hotel. For a 35-hour week she makes a gross salary of $477.75. What is her hourly wage?

6. David makes $17.90 an hour installing drywall. Calculate the amount of F.I.C.A. he has to contribute in a week when he works 35 hours.

7. Patricia works as a legal assistant. Her gross weekly income is $788.40. According to the tax withholding tables on page 15, how much federal income tax is withheld from her check each week if she is single and claims four withholding allowances?

8. Sam is also a legal assistant, and he too makes $788.40 a week. How much federal income tax is withheld from his check each week if he is married and claims three withholding allowances?

9. Eve is a real estate agent. She earns a commission of 6% on every sale. What was her commission on a house that sold for $159,500?

10. Jorge works as a car salesman. He gets a commission of 3.5% on used cars and 5.4% on new cars. What was his commission for a month when he sold used cars for a total of $16,583 and new cars for a total of $34,792?

Part B

Use the following information to answer problems 11 to 14.

SITUATION

> Adrienne is a physical trainer. The health club where she works pays her a yearly salary of $42,680.

11. What is her gross weekly income?

12. How much is her weekly F.I.C.A. contribution?

13. How much is withheld from her weekly paycheck for federal income tax if she is single and claims two withholding allowances?

14. What is Adrienne's net weekly income if she contributes $37.50 each week to her retirement account?

Part C

Use the following information to answer problems 15 to 18.

SITUATION

> Manny is a carpenter, and he makes $23.80 an hour. He is married and claims five withholding allowances.

15. What is his gross weekly income for a 40-hour week?

16. What is his weekly F.I.C.A. contribution?

17. How much is withheld from his weekly paycheck for federal income tax?

18. What is Manny's net weekly income if he pays $12.50 every week in union dues?

Part D

Use the following information to answer problems 19 to 22.

SITUATION

Miles heads the accounting department for a mail-order catalog business. His yearly salary is $52,380. He is married and claims three withholding allowances.

19. What is his gross weekly income?

20. What is his weekly F.I.C.A. contribution?

21. How much is withheld from his weekly paycheck for federal income tax?

22. What is Miles' net weekly income if he contributes $45 each week to his retirement account?

Post-Lesson Vocabulary Reinforcement

Choose the appropriate term to complete each sentence.

1. _____, 12.4% of every employee's income, up to an annual limit, must be contributed to Social Security.

2. The overtime _____ is often _____ a worker's regular hourly wage.

3. Deductions from a worker's _____ may include Social Security, Medicare payments, federal income tax, state income tax, contributions to retirement programs, _____, and _____ contributions.

4. An employer withholds more money from a worker who claims 0 _____ than from a worker who claims 1 or more.

5. The Internal Revenue Service sets _____ to determine the amount an employer withholds from a worker's paycheck.

6. To _____ how much Simon's gross weekly income is if he _____ $12.80 an hour, you have to know how to multiply _____.

7. According to the Federal Insurance Contributions Act, or F.I.C.A., 2.9 _____ an employee's income must be contributed to Medicare.

Match the definitions with the terms.

_____ 8. the branch of the federal government that collects taxes

_____ 9. an organization that supports workers and helps to protect their rights

_____ 10. money paid to the federal government for its operation

_____ 11. income per year

_____ 12. an amount of money a person gets or earns, usually from working at a job

_____ 13. income per (for an) hour

_____ 14. money paid to the state government for its operation

_____ 15. income deductions used to pay for medical care at the age of 65 and beyond

_____ 16. extra, more than the regular amount of time worked

_____ 17. money subtracted from income and paid out when one retires

charitable

withholding allowances

earns

paycheck

union dues

find

rate

$1\frac{1}{2}$ times

decimals

% of

by law

guidelines

a. Social Security deductions

b. Medicare payments

c. federal income tax

d. state income tax

e. Internal Revenue Service

f. a union

g. income

h. hourly wage

i. yearly salary

j. overtime

Language Builder

ACTIVITY A The **simple** or **command** form of the verb is used frequently in this lesson and throughout this book. Read over the lesson and find command forms. Add them to the list below.

1. Multiply, Find, _____ , _____ , _____

ACTIVITY B Read the sentences about people who do the wrong thing. Tell them to do the right thing. Use the words in parentheses and the command form of the verb to write a new sentence.

2. John multiplies his overtime rate by the number of hours he works each week. (your hourly wage . . . you)

    Multiply your hourly wage by the number of hours you work each week.

3. Simon assumes that a year has 53 weeks. (52 weeks)

4. John divides Sara's salary by 50. (by 52)

5. The student looks at the second table. (the first table)

ACTIVITY C The **simple present** and **past tense** forms of the verb are also used frequently in this lesson and throughout this book. Change present tense forms to past tense forms. You can check the lesson for the past forms.

6. work   *worked*

7. contribute   _____

8. withhold   _____

9. claim   _____

10. sell   _____

ACTIVITY D On a separate sheet of paper, change the following sentences from the **present** to the **past tense.** Then find past tense sentences in the lesson that almost match the ones you wrote.

11. How much does John make in overtime wages in a week when he works 46 hours?

12. John claims three withholding allowances on his W-4 form.

13. What is Eve's commission on a house that sells for $159,500?

LESSON 2: BANKING

Pre-Lesson Vocabulary Practice

Read the words and their meanings on the right. Next carefully read the lesson, and try to find the words. These words appear in boldfaced type in the lesson. Then write sentences using these words and share them with a partner.

Example: *I stopped at an ATM to get money for the movie.*

Below is a list of words that appear throughout the lesson. Read each word and its definition. Then work with a partner to ask and answer questions using the words.

convenience – ease, something not difficult

to write a check – to fill in, or write down, payment information on a blank check

a bank account – money placed in a bank

a withdrawal – an amount of money removed from, or taken out of, a bank account

at one time – in the past, previously

have become less popular – are not as common as before

interest – money a customer earns that represents a percent of the balance in an account

branches – banks that belong to the same main bank, but are in different locations

savings account – a customer's bank deposits that earn special interest over a period of time

checking account – a customer's bank deposits used for writing checks

to charge (for a transaction) – to have a customer pay for a specific transaction or service

a statement – a report that shows the separate transactions in an account

a challenge – a test, an experience that might be difficult

telephone transfers – money moved from one account to another by making a telephone call

to bring (something) up-to-date – to make it current—for example, to correct a transaction register

ATM – a machine used to make transactions

to balance (reconcile) – to compare a transaction register with a bank statement

checkbook – a booklet that contains checks and a transaction register

cleared (as in checks) – checks that have been paid and cashed

credit – an amount earned

debit – an amount subtracted from a bank balance

debit card – a card used in place of a check

deposit – to put money into a bank account

interest rate – a percent that may change over time

minimum balance – money the bank requires in an account

modest interest – a small amount of money earned

service charges – money a customer pays a bank for transactions

transaction – an action or operation

transaction register (passbook) – a booklet used to record bank transactions

withdraw – to take money out of a bank account

Banking

People use banks for the convenience of having a place to **deposit** paychecks, to be able to write checks, and to **withdraw** cash from **ATMs**—automated teller machines. At one time, many people had **passbook** savings accounts in banks. A passbook is a record of deposits, withdrawals, and earned interest. As **interest rates** have dropped, passbook savings accounts have become less popular. (You will learn more about interest in Lesson 9.)

Banks charge for their services in a variety of ways. Some banks charge for every **transaction.** Other banks offer free services such as checking as long as the customer maintains a **minimum balance.** Banks often pay a modest interest on the balance that is kept in an account. Most banks do not charge their own customers for withdrawals from the ATMs at their branches. However, most banks do charge for withdrawals if a customer from another financial institution uses the bank's machines.

Every month banks send each checking account customer a **statement** that shows the details of the activity in the account. The statement lists deposits, checks that have **cleared,** earned interest, **service charges,** withdrawals from ATMs, and the use of **debit cards.** A debit card is a plastic card that can be used instead of writing checks. When a customer presents a debit card at a store, the amount of the purchase is deducted from her checking account.

The challenge for a banking customer is to compare personal financial records with the bank's monthly statement. Every **checkbook** comes with a **transaction register** to write down each **debit** or **credit.** To **balance** a checkbook, a customer should add every credit to the existing balance and subtract every debit. Sometimes checkbooks don't balance, and the customer will have to contact the bank to solve the problem.

Comparing personal financial records with the monthly statement from a bank is called **reconciling.** A customer's checkbook register may list checks that were written but have not yet cleared the bank when the monthly statement is printed. The mathematics of reconciling is simple:

A customer should add:
 deposits
 interest earned

A customer should subtract:
 checks paid
 ATM withdrawals
 telephone transfers and payments
 service charges
 debit card purchases

To reconcile a checking account with a bank statement, first bring the checkbook register up to date by adding any interest earned and subtracting any service charges.

Then bring the bank statement up to date by adding late deposits, subtracting checks that have not cleared, and subtracting unlisted withdrawals from ATMs. The two corrected balances should agree.

EXAMPLE At the end of June, Tim's checkbook balance was $619.22. His banking statement for June showed a balance of $639.57. The statement included an interest payment of $0.63 and a service charge of $2.75. The bank statement did not include a check for $14.63 to a hardware store, a check for $83.89 to a fuel oil company, an ATM withdrawal for $140.00, a debit card purchase of $20.15, or a deposit of $236.20.

Solution To bring Tim's checkbook records up-to-date, add the interest and subtract the service charge.

$619.22 + $0.63 - $2.75 = **$617.10**

To bring the bank statement up-to-date, add the late deposit, subtract the two checks that have not cleared, subtract the ATM withdrawal, and subtract the debit card purchase.

$639.57 + $236.20 - $14.63 - $83.89 - $140.00 - $20.15 = **$617.10**

Since the corrected balances are the same, the checkbook and the bank statement are reconciled.

To solve the problems in the next exercise, review:
- adding and subtracting decimals, page 230

EXERCISE 2

Part A

Use the following information to answer problems 1 to 5.

SITUATION

> At the end of February, Lupé's checkbook register showed a
> balance of $371.49. Her February bank statement showed a balance
> of $357.97. The bank statement included $0.26 interest and a service
> charge of $2.50. The statement did not include an ATM withdrawal
> of $60.00 that she made on February 26, and it did not show a check
> for $53.72 that she wrote on February 25. A deposit that she made
> on February 24 of $125.00 was not listed on the bank statement.

1. Correct Lupé's checkbook register for February by adding the interest she earned and subtracting the service charge.

2. To correct the bank statement, first add any unrecorded deposits to the total on the statement.

3. What is the total of the checks and withdrawals that do not appear on the bank statement?

4. Subtract the answer to problem 3 from the total in problem 2.

5. Is the checkbook balance reconciled with the bank statement?

Part B

Use the following information to answer problems 6 to 8. Then fill in the blank spaces in the checkbook register.

SITUATION

At the end of April, Keisha's checkbook register had a balance of $763.28. Her April bank statement showed a balance of $902.96. The statement included $0.27 interest and a service charge of $3.25 from April 29. The statement did not include an ATM withdrawal of $140 that she made on April 27 or an ATM withdrawal of $80.00 that she made the next day. The statement did not list a check for $19.53 that Keisha wrote on April 23 or a check for $112.96 that she wrote on April 24. A deposit that Keisha made on April 23 for $209.83 was not listed on the bank statement.

6. To correct Keisha's April checkbook register, add the interest that she earned for the month and subtract the service charge.

7. Add the unlisted deposit to the total on the bank statement.

8. Find the total of the checks and withdrawals that do not appear on the bank statement.

NUMBER	DATE	DESCRIPTION OF TRANSACTION	PAYMENT/DEBIT(-)		DEPOSIT/CREDIT(+)		$	905.94
112	4/23	check	$ 19	53	$			
	4/23	deposit			209	83		
113	4/24	check						
	4/27	ATM						
	4/28	ATM						
	4/29	interest						
	4/29	service charge						

Part C

Below is the September page from Miguel's checkbook register and his September bank statement. Use the register and the bank statement to answer problems 9 to 13.

NUMBER	DATE	DESCRIPTION OF TRANSACTION	PAYMENT/DEBIT(-)		DEPOSIT/CREDIT(+)		$	708.16	
533	9/3	Carl's Car Repair	$ 83	05	$			83	05
								625	11
534	9/5	Atlantic Telephone	44	61				44	61
								580	50
	9/10	ATM	200	—				200	—
								380	50
535	9/12	Niagara Utilities	33	52				33	52
								346	98
536	9/16	Capital Mastercard	150	—				150	—
								196	98
	9/24	deposit (paycheck)			614	96		614	96
								811	94
	9/25	ATM	100	—				100	—
								711	94
537	9/28	Hal's Hardware	49	98				49	98
								661	96

9. To correct Miguel's September checkbook register, add the interest that he earned for the month and subtract the service charge.

10. Add any unlisted deposits for September to the ending balance on the bank statement.

11. What is the total of the checks and withdrawals that do not appear on the bank statement?

12. Subtract the answer to problem 11 from the new total in problem 10.

13. Are the checkbook register and the bank statement reconciled?

September Bank Statement

Withdrawals and Debits

Date	Description	Amount
9/10	ATM withdrawal	$200.00
8/11	ATM service charge	$4.00

Checks Paid

Check	Date	Amount
533	9/12	$83.05
534	9/15	$44.61
535	9/22	$33.52
536	9/12	$150.00

Deposits and Credits

Date	Description	Amount
9/20	Interest Credit	$0.43

Ending Balance
$193.41

Part D

14. Use the information for check number 537 in Miguel's check register to fill out the blank check below.

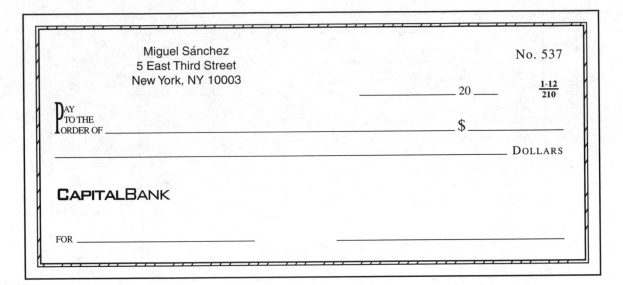

Miguel Sánchez
5 East Third Street
New York, NY 10003

No. 537

_____ 20 ____ $\frac{1\text{-}12}{210}$

PAY
TO THE
ORDER OF _____ $ _____

_____ DOLLARS

CAPITALBANK

FOR _____ _____

Post-Lesson Vocabulary Reinforcement

Look at the bank statement. In the blanks, write the appropriate letters from the list below.

<table>
<tr><td colspan="4">

September Bank Statement

Beginning Balance　　$575.00

Withdrawals and Debits

Date	Description	Amount	
9/11	ATM withdrawal	$200.00	❶ _____
8/11	ATM service charge	$4.00	❷ _____
8/10	Debit card purchase	$14.58	❸ _____

Checks Paid　　　　　　　　　　❹ _____

Check	Date	Amount
533	9/12	$83.05
534	9/15	$44.61
535	9/22	$33.52
536	9/12	$150.00

Deposits and Credits　　　　　❺ _____

Date	Description	Amount	
9/20	Interest Credit	$1.43	❻ _____
9/21	Check Deposited	$198.45	❼ _____

Ending Balance　　　　　　　　❽ _____
$245.12

</td></tr>
</table>

a. the amount of a purchase made with a debit card

b. the total money in the bank after all transactions have been reconciled

c. money taken from a bank account via an automated teller machine

d. an amount of money put into a bank account from a check

e. money taken from an account to pay for an ATM transaction

f. money earned based on the balance in the account

g. money put into a bank account and money earned based on the balance in an account

h. checks that have cleared

Match the sentences with the same meanings. Write the appropriate letters in the blanks.

a. Banks charge for their services in a variety of ways.

b. Most banks charge for ATM withdrawals for people who are not customers.

c. A debit card is a plastic card that can be used instead of writing checks.

d. A checkbook doesn't balance.

e. When reconciling an account, add to or subtract from the checkbook register any unlisted transactions on the bank statement.

_____ 9. A transaction register is not the same as a bank statement.

_____ 10. Some banks have service charges that other banks don't have.

_____ 11. If a person uses an ATM at a bank that is not his or her own, there is a charge for the service.

_____ 12. A person has to compare the transaction register with the bank statement to bring the register up-to-date.

_____ 13. This is a card that is used in place of a check.

Language Builder

ACTIVITY A The **infinitive of purpose** is used several times in this lesson, followed by the **command form** (to + simple form of the verb) of the verb (which you practiced in Lesson 1). Here is an example from the lesson: **To correct** Miguel's September checkbook register, **add** the interest that he earned for the month and **subtract** the service charge.

1. There are six more examples of this kind of sentence in the lesson. Find them and write them on a separate sheet of paper.

ACTIVITY B Match the command forms with the infinitives of purpose to complete each sentence. Write the correct letters in the blanks.

_____ 2. To pay Niagara Utilities,

_____ 3. To put money in your bank account,

_____ 4. To take money from your account,

_____ 5. To solve a problem of a checkbook that won't balance,

a. make a withdrawal.

b. make a deposit.

c. write a check.

d. contact your bank.

ACTIVITY C **Gerunds** are nouns that end in **-*ing***. A gerund can appear at different places in a sentence, as in this sentence from the lesson: **Comparing** personal financial records with the monthly statement from a bank is called **reconciling.** Check (✓) the sentences that have gerunds.

_____ 6. People use banks for the convenience of having a place to deposit paychecks.

_____ 7. With a debit card, the purchase is deducted from the checking account.

_____ 8. Some banks offer free services, such as checking, as long as the customer maintains a minimum balance.

ACTIVITY D Write the correct gerund from the box in the appropriate blank. Some gerunds should be used twice.

balancing	**adding**	**reconciling**	**subtracting**

9. The mathematics of _____ is simple: Bring your bank statement up-to-date by _____ late deposits, _____ checks that have not cleared, and _____ unlisted withdrawals from ATMs.

10. _____ a checkbook means _____ every credit to the existing balance and _____ every debit.

Pre-Lesson Vocabulary Practice

Match the terms on the right with the correct pictures below.

a. a calculation

b. a calculator

c. a coupon

d. price sticker from a store shelf

e. list price of an item

f. discounted item

1. _____

2. _____

3. _____

4. _____

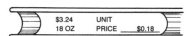

5. _____

6. _____

Study the vocabulary items below. Then find them in the lesson.

combined – added together

end-of-season sale – items with discounted prices because a particular season is over

list price – the original cost of an item

maximum – the greatest amount; the opposite of minimum

on top of – in addition to

original – the initial cost before deductions or additions

a purchase – something paid for

raises – increases

sales tax – money charged by a government and added to the original price of an item

shortcut – an easier and quicker way to do something

subtotal – a total amount before another amount is added or subtracted

to save – to not spend

to suppose – to assume

to take advantage of – to use and get the benefit of

to the nearest dollar – to make a rounding calculation

to vary from state to state – to change, depending on the state

worth – a dollar amount

Work with a partner. One partner should read the terms from the list above, but in a different order. The other should say the meanings. Then take turns—one of you reading aloud a term and the other finding and reading aloud the correct meaning.

Shopping

Today consumers shop over the telephone and on the Internet as well as at stores. But the mathematics of shopping is the same as it has always been. The amount a customer pays is the total of the **list prices** for the items that the customer purchases.

In a department store, the list price may be on a sticker attached to the item. In a grocery store, the list price may be written on a tag on the shelf below the item. The list price may appear beside the description of the item in a catalog or on a company's Web page.

Use the cost formula to find the price of an item. The cost formula is $c = nr$ where c is the cost of an item, n is the number of items, and r is the rate or cost for a single item.

> **EXAMPLE 1** Irma bought 2.3 pounds of beef that cost $2.49 per pound and 1.7 pounds of chicken that cost $1.19 per pound. What was the total cost of her purchases?

Solution **Step 1.** Find the cost of the beef.

$$2.3 \times \$2.49 = \$5.727 \text{ or } \$5.73$$

Step 2. Find the cost of the chicken.

$$1.7 \times \$1.19 = \$2.023 \text{ or } \$2.02$$

Step 3. Add the two costs.

$$\$5.73 + \$2.02 = \mathbf{\$7.75}$$

In many states a customer also has to pay a **sales tax.** Sales tax is a percent of the **subtotal** of the items a customer purchases.

EXAMPLE 2 Michelle bought a bottle of shampoo for $1.89, a container of detergent for $4.99, and a tube of toothpaste for $1.29. What was the total of these items if the sales tax rate in Michelle's community is 6.5%?

Solution

Step 1. To find the subtotal, add the prices of the items Michelle bought.

$1.89 + $4.99 + $1.29 = $8.17

Step 2. Find 6.5% of the subtotal: 6.5% = 0.065.

0.065 × $8.17 = $0.53105 or $0.53

Step 3. Add the subtotal and the sales tax.

$8.17 + $0.53 = **$8.70**

The last two steps can be combined in a shortcut. Think of the subtotal as 100%. The total price including sales tax is 100% + 6.5% = 106.5%. The total cost of Michelle's purchases is

1.065 × $8.17 = $8.70105 or **$8.70**

Sales tax rates vary from state to state and even from town to town. Communities sometimes add an additional tax on top of the state sales tax rate. Also, sales tax may not apply to all items. Some items such as food or clothes, if they are below a certain price, may be exempt from sales tax. The word *exempt* means "free from" or "not subject to."

A sales tax raises the total price a customer pays. However, when an item is "on sale," a customer gets a **discount.** A discount means that the list price is reduced. The discount is often a **percent off** the original price. The sale price is the list price minus the discount.

EXAMPLE 3

What is the sale price of a pair of jeans listed at $24.89 but on sale for 20% off the list price?

Solution **Step 1.** Find 20% of the list price: 20% = 0.2.

0.2 × $24.89 = $4.978 or $4.98

Step 2. Subtract the discount from the list price.

$24.89 − $4.98 = **$19.91**

Notice the difference between the two simple words *off* and *of*. The term "20% off" the list price means to subtract "20% of" the list price from the list price. The word *off* suggests subtraction, and the word *of* suggests multiplication.

There is a shortcut to finding a percent off a list price. Again, think of the original price as 100%. 20% off means 100% − 20% = 80%. The sale price is 80% of the original price. For the jeans in Example 3, the sale price is:

80% of $24.89 = 0.8 × $24.89 = $19.912 or **$19.91**

Another way to reduce the price of items is with a **coupon.** When a customer presents the coupon to a cashier, the price of an item is reduced or **discounted.**

To solve the problems in the next exercise, review:

- multiplying decimals, page 231

- finding a percent of a number, page 236

- finding what percent one number is of another, page 238

EXERCISE 3

Part A

Use a calculator to solve the problems in this exercise.

Meat		Dairy	
Ground Beef	$1.99/lb	Jarlsberg Cheese	$7.19/lb
Rib Steaks	$5.99/lb	Milk	$1.49/half gallon
Pork Chops	$2.49/lb	Sour Cream	$1.29/16-oz carton
Turkey Breast	$1.49/lb	Yogurt	$3.19/32-oz container

Fruits and Vegetables		Household Items	
Cucumbers	2 for $1	Dishwashing Soap	$1.69/32-oz bottle
Green Beans	$0.99/lb	Laundry Detergent	$4.99/80-oz liquid
Melons	$0.89/lb	Shampoo	$1.79/8-oz bottle
Peaches	$1.09/lb	Toothpaste	$1.08/12-oz tube

Use the price list above to answer problems 1 to 5. Notice the slash (/) with many of the prices. The slash represents the word *per*. For example, ground beef costs $1.99 per pound or $1.99/lb.

1. Rick bought 1.8 pounds of ground beef and 1.3 pounds of green beans. Assuming there is no sales tax on these items, what was the cost of Rick's purchases?

2. Marlene bought 2.4 pounds of melons, a 32-ounce bottle of dishwashing soap, and a 32-ounce container of yogurt. If none of the items had sales tax, what was the total of her purchases?

3. Mr. Ryan bought 3.65 pounds of peaches and an 80-ounce container of liquid laundry detergent. If there is a 5% sales tax on non-food items, what was the total of these purchases?

4. Nate bought a piece of Jarlsberg cheese that weighed 0.45 pound and 2.1 pounds of turkey breast. If there was no sales tax on these items, what was the total of his purchases?

5. Deborah bought two half-gallon containers of milk, four cucumbers, and an 8-ounce bottle of shampoo. She presented a coupon for $0.50 off the price of the shampoo to the cashier. If there is a 6% sales tax on all of these items, what is the total of her purchases including tax?

Part B

The table below compares state and local sales tax rates in five states. Use the table to answer problems 6 to 10.

Sample State and Local Retail Sales Taxes				
	Food taxable (T) exempt (E)	State Rate	Maximum Local Rate	Maximum State/Local Rate
Arizona	E	5.60%	3.00%	8.60%
California	E	6.00	2.50	8.50
Missouri	T	4.225	4.125	8.35
New Jersey	E	6.00	--	6.00
Utah	T	4.75	2.25	7.00

6. Which of the states shown on the table has the highest maximum combined state and local tax rate?

7. Which state shown on the table has the lowest basic state tax rate?

8. In New Jersey, what is the total cost including sales tax of a shirt that is listed for $29.98?

9. A chain saw is listed for $154.95. What is the cost of the saw if it is sold at a store in a community in Arizona that has no local sales tax?

10. Suppose the total for food items at a grocery store is $48.27. Compare the total, including sales tax, if the groceries were purchased in a California town with the maximum state and local tax rate and in a grocery store in a Missouri town with the maximum state and local tax rate.

Part C

11. A vacuum cleaner listed for $399 is on sale at 10% off the list price. What is the total price of the vacuum cleaner if it is sold in a community where the sales tax rate is 6%?

12. If an item is on sale for 15% off the list price, a customer will pay what percent of the list price before tax?

13. At an end-of-the-season sale, winter coats are marked 40% off. Find the sale price of a coat originally selling for $139 if the sales tax is 7.5%.

14. A portable radio and CD player had a list price of $199. If it is now on sale for 10% off the list price, what is the total cost in a community that charges a sales tax of 8.35%?

15. To the nearest dollar, how much less is the sale price, including tax, of the radio and CD player in problem 14 than the list price?

Part D

Use the following table to answer problems 16 to 19.

At Phil's Furniture,	
Get	**When you spend**
$50	$250 to $499
$75	$500 to $999
$150	$1,000 to $1,999
$300	$2,000 to $2,999
$450	$3,000 or more

16. If an item at Phil's Furniture is listed at $250, what percent of the price does a customer save?

17. A sofa and matching armchair are listed at $1,999 at Phil's Furniture. Approximately what percent of the list price can a customer save?

18. A reclining chair is listed at $299 at Phil's. What is the sale price of the chair?

19. Sonia bought $2,460 worth of furniture from Phil's Internet site. If she takes advantage of the sale offer, how much will the furniture cost including 5% sales tax?

Post-Lesson Vocabulary Reinforcement

Choose the appropriate term from the list on the right to complete each sentence.

39

1. Another word for customer is _____.

2. _____ is a percent of the subtotal of the items a customer purchases.

3. When an item is "on sale," the customer gets a _____.

4. _____ is the list price minus the discount.

5. When a customer presents a _____ to a cashier, the price of the item is discounted.

6. In some places, items such as food are _____ from sales tax if they are below a certain _____.

7. Communities sometimes _____ an additional tax _____ the state sales tax rate.

8. At the _____ sale, winter coats are marked 40% _____.

9. Sal and Sonia will take advantage of the furniture sale and _____ a lot of money.

sales tax

exempt

sale price

discount

price

on top of

coupon

consumer

add

end-of-season

off

save

Look at the bill. In each blank, write the appropriate letter from the list below.

10. _____

11. _____

12. _____

13. _____

14. _____

$399.00

−39.90 (10% off)

$359.10

+ 29.99 (8.35% Sales Tax)

$389.09

a. a discounted amount

b. the subtotal

c. the original price

d. the total price

e. an amount on top of (added to) the subtotal

Language Builder

ACTIVITY A The **passive form** of the verb is used several times in this lesson. In fact, the sentence you just read uses the passive! Here's an example from this lesson:

The saw is sold in a community that has no local taxes.

Description: The **passive form** of the verb in this lesson <u>is</u> <u>used</u> when it <u>is</u> not <u>known</u> who performed the action of the verb because that information is not important. The passive occurs frequently in English. This paragraph has three passives! They <u>are</u> <u>underlined</u> for you.

What makes a verb passive? The **verb to be** (*is / are*) + the **past participle** of the verb (*used,* for example) makes the verb passive. Below write the other past participles from the passives in the above paragraphs:

1. _____   _____   _____

ACTIVITY B Below are some sentences similar to those in this lesson. On a separate sheet of paper, change the verb forms in parentheses ( ) and rewrite the sentences in the passive. Use the present tense. The first sentence is completed as an example.

2. A discount means that the list price (*be / reduce*).

   A discount means that the list price is reduced.

3. What is the sale price of a shirt that (*be / list*) at $24.89 less 20%?

4. At an end-of-season sale, winter coats (*be / mark*) 40% off.

5. A reclining chair (*be / list*) at $299 at Phil's Furniture.

6. A sofa and matching armchair (*be / list*) at $1,999 at Phil's.

ACTIVITY C A **superlative adjective** (a word that may end in *-est*) means that the noun that follows it is the maximum or minimum. For the following phrases write *maximum* in the blank if the superlative adjective means "the greatest amount" or write *minimum* if the superlative adjective means "the least amount."

7. _____ the lowest basic state tax rate

8. _____ the highest tax rate

Now look at the following sentences. Underline the *-est* words that are superlative adjectives.

9. Which state has the highest tax rate?

10. Calculate the difference between the list price and the sale price to the nearest dollar.

LESSON 4: UNIT PRICING

Pre-Lesson Vocabulary Practice

Read the words and their meanings on the right. Then carefully read the lesson, and try to find the words.

Work with a partner. Quiz each other on the vocabulary words and their meanings. Then take turns—one of you reading aloud a term from the list while the other finds the correct meaning and reads it aloud.

Read the definitions of the units of measurement.

ounce – a small liquid measurement, equal to $\frac{1}{16}$ of a pint

quart – 2 pints

pint – 16 fluid ounces

gallon – 4 quarts

Now write the appropriate names for the different units from largest to smallest in the blanks below.

1. _____

2. _____

3. _____

4. _____

area – total surface of length × width

better value / better buy – a more economical purchase than another purchase; a lower price

high-nutrition – good for one's health

linear unit – a length, such as an inch or a foot

liquid – a fluid such as water

packaged – the way or manner in which something is contained

pound – a weight of sixteen ounces

quickly – rapidly, fast

regular grade gasoline – the lowest-priced liquid used to power vehicles

required – necessary

to compare – to see how two or more things are similar or different

to display – to show

to measure – to find a specific dimension or quantity

to round – to make an estimate that is close to an original amount

Unit Pricing

To shop wisely, a customer should **compare** prices of similar items. However, it is not always easy to compare prices when similar items are packaged differently.

For example, which is a better value, a 10-ounce can of string beans that costs $0.59 or a 16-ounce can of string beans that costs $1.04?

Most grocery stores are now required by state laws to display the **unit price** of the items on their shelves. A unit price is the price for one unit of measurement such as an ounce or a quart.

To calculate a unit price, divide the cost of an item by the number of units in the item.

A unit price is the cost *per* unit. The word *per* is a Latin word that means **for each.** Use a calculator to find unit prices quickly.

EXAMPLE 1   Which of the two cans of string beans is a better buy?

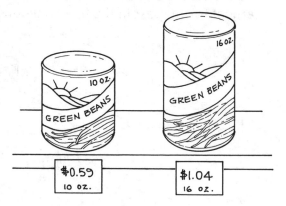

Solution   The unit price of the 10-ounce can is

$0.59 ÷ 10 = $0.059 per ounce.

The unit price of the 16-ounce can is

$1.04 ÷ 16 = $0.065 per ounce.

The 10-ounce can is the better buy because the unit price is lower than the unit price for the 16-ounce can.

Unit pricing is based on **weight, liquid measure, count,** or **area.** The unit prices of the cans of beans in the last example were measured in the price per unit of weight. Meat, poultry, fish, and frozen foods are usually measured in units of weight.

Read the labels on packaged items carefully. A label that gives the price for 2.4 pounds of chicken is not giving the price for 2 pounds 4 ounces of chicken. The price is for two and four-tenths pounds.

The unit prices of beverages, oils, and cleaners are usually given as the price per liquid unit such as a pint, a quart, or a fluid ounce.

EXAMPLE 2   Find the unit price of a 13.5-fluid ounce bottle of shampoo that costs $2.19.

Solution   **Divide to find the unit price.**

$2.19 ÷ 13.5 = $0.16222. . . or about **$0.16 per fl oz**

The price of items such as scrubbing pads, tea bags, and facial tissues is often measured in the price per count. The unit price is the price for one item in the package. To find the unit price, divide the price of the package by the number of items in the package.

EXAMPLE 3 What is the unit price of 100 paper napkins that cost $1.99?

Solution   Divide to find the unit price.

$1.99 ÷ 100 = $0.0199 or about **$0.02 per napkin**

The unit price of some products such as wax paper, aluminum foil, and paper towels is given as the price for a unit of area such as a square foot or a square meter. The formula for the area of a rectangle is $A = lw$ where A is the area, l is the length, and w is the width. Area is measured in square units such as square feet, square inches, or square meters. Length and width are measured in linear units such as feet, inches, or meters.

EXAMPLE 4 What is the area in square feet of the aluminum foil in a roll that is 12 inches wide and $8\frac{1}{3}$ yards long?

Solution   The length is $8\frac{1}{3} \times 3 = \frac{25}{3} \times \frac{3}{1} = 25$ feet.

The width is 12 inches or 1 foot.

$A = 25 \times 1 =$ **25 square feet**

EXAMPLE 5 Find the unit price of a 25-square-foot roll of aluminum foil that costs $1.15.

Solution   The unit price is $1.15 ÷ 25 = $0.046 or about **5 cents per square foot.**

To solve the problems in the next exercise, review:

- dividing decimals, page 231
- units of measurement, page 239
- rounding decimals, page 229
- calculating area, page 241

EXERCISE 4

Part A

Use a calculator to solve these problems.

1. To the nearest penny, what is the unit price of a package of ground beef that weighs 2.6 pounds and costs $4.13?

2. A plastic container holds 180 multiple vitamins. Find the unit cost if the price of the container is $12.49.

3. A half gallon of milk costs $1.79. What is the cost per quart of the milk?

4. What is the cost per gallon of the milk in problem 3?

5. What is the unit cost of a 100-square-foot roll of aluminum foil that costs $3.29?

6. What is the area in square feet of the wax paper in a roll that is 12 inches wide and $66\frac{2}{3}$ yards long?

7. A 75-square-foot roll of Ace aluminum foil costs $2.79. A 25-square-foot roll of Baxter aluminum foil costs $0.99. Which is the better buy?

8. Charlotte wants to get the best value among these three brands of aspirin. Which brand is the best buy?

   Brand A costs $1.99 for 36 aspirin.

   Brand B contains 100 aspirin and costs $5 for 2 jars.

   Brand C costs $2.69 and contains 100 aspirin.

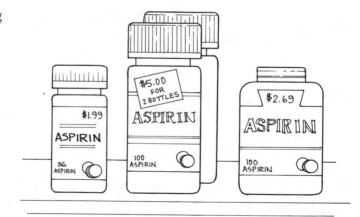

9. Breakfast cereals come in a wide variety of packaging and prices. Paulo is comparing the prices of three high-nutrition breakfast cereals. A box of Healthy Start cereal contains 12.5 ounces and costs $3.29. A box of Better Body cereal costs $3.79 for 11.4 ounces. A box of Good Grains cereal costs $3.19 for 10.75 ounces. Which is the best buy?

Part B

Use the table to answer problems 10 to 15.

Cost of Filling an 18.5-Gallon Tank with Regular Grade Gasoline January 2011		
Region	**Total Cost**	**Unit Price**
U.S. Average	$57.25	
East Coast	$57.45	
New England	$58.69	
Central Atlantic	$58.19	
Lower Atlantic	$56.40	
Midwest	$57.02	
Gulf Coast	$54.46	
Rocky Mountains	$53.47	
West Coast	$60.59	

10. In which region is regular grade gasoline most expensive?

11. In which region is regular grade gasoline least expensive?

12. For which region shown in the table is the price of regular grade gasoline closest to the U.S. average?

13. Find the unit price to the nearest tenth of a cent for each region shown on the table. (**Hint:** To find the cost to the nearest tenth of a penny, round each answer to the third decimal place, or thousandths.)

14. To the nearest penny, how much more does it cost to buy 10 gallons of regular grade gasoline at the average Gulf Coast price than 10 gallons at the average Rocky Mountain price?

15. The average price of a gallon of diesel fuel in January 2011 was $3.388. How much more is the price than the U.S. average price for a gallon of regular grade gasoline?

Post-Lesson Vocabulary Reinforcement

Look at the pictures of the different items. Choose the appropriate name of each item from the box below and write it on the first blank under the item.

| string beans | shampoo |
| aluminum foil | napkins |

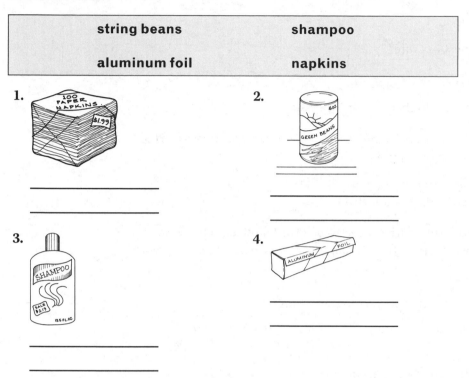

1.

_____

_____

2.

_____

_____

3.

_____

_____

4.

_____

_____

area

liquid

weight

count

Now decide which unit of measurement each item above is based on. Choose the appropriate words from the list on the right and write them on the second of the lines above.

Choose the appropriate term from the list on the right to complete the sentences.

5. To shop _____, a customer should _____ prices of similar items.

6. Unit pricing is based on _____, liquid measure, count, or area.

7. A _____ is the price for one unit of measurement such as an ounce or a quart.

8. Most grocery stores are now _____ by state laws _____ the unit price of the items on their shelves.

9. To calculate a _____ price, divide the cost of an item by the number of units in the item.

10. The string beans are _____ in different-sized cans.

11. The 10-ounce can is the _____ because the price per ounce is lowest.

weight

unit price

compare

to display

packaged

required

wisely

better buy

unit

Language Builder

ACTIVITY A In Lesson 3, you practiced recognizing and using the **superlative adjective** (a word that may end in -*est*). **Comparative adjectives** (words that may end in -*er*) have a similar function, except they are used to compare two things. Read the following sentences and underline the superlative and comparative adjective.

Superlative Adjective

1. Which state has the highest tax rate?

Comparative Adjective

2. Sales taxes in some towns are higher than in others.

   Notice that the **regular form** of comparative and superlative adjectives is formed by adding -*er* or -*est* to the base adjective. For example:

> low + er = lower (comparative adjective)
> low + est = lowest (superlative adjective)

Underline the comparative and superlative adjectives in the following sentences. Remember to underline only adjectives (words that modify nouns), not just any word that ends in -*er* or -*est*.

3. A shortcut is an easier and quicker way to do something.

4. Find the unit price to the nearest tenth of a cent for each region shown on the table.

5. The "best value" means the greatest value of three or more items.

ACTIVITY B **Adverbs** are words that can modify verbs. Adverbs frequently end in the letters -*ly*. The sentence you just read has an adverb. Underline it below. Then double underline the verb that it modifies.

Adverbs frequently end in the letters -ly.

Now compare your work with the answer below:

Adverbs <u>frequently</u> <u>end</u> in the letters -*ly*.

Adverbs can occur in different places in a sentence—sometimes before the verb, sometimes immediately after the verb, and sometimes in other places. Underline the adverbs in the following sentences. Then double underline the modified verbs.

6. To shop wisely, a customer should compare prices of similar items.

7. Use a calculator to find unit prices quickly.

8. Read the labels on packaged items carefully.

ACTIVITY C Use **comparative** and **superlative adjectives** and **adverbs** from this lesson in your own sentences. Then share and compare your work with a partner and with the class.

LESSON 5: DINING OUT

Pre-Lesson Vocabulary Practice

Read the words on the right and their meanings. Next find them in the lesson—they are in boldface. Then carefully read the lesson material.

Using the words on the right, complete the sentences below.

1. The bill shows the _____ of all the items we ordered.

2. _____ is added to the total cost of the items we ordered.

3. My friend paid the entire _____.

4. We decided to leave a 15% _____.

5. The amount given to a cashier is the _____ and the change you get back is the _____.

Below is a list of words that appear throughout the lesson. Read each word and its definition. Then work with a partner to use the words in a sentence.

adequate – enough, sufficient

altogether – in total, considering everything

as well as – also, in addition to

average – standard, regular, mean

before-tax total or **pre-tax total** – the total cost before adding the tax

calculate – use numbers to determine or figure something out

common standard – the usual or acceptable amount

cover – pay completely

eat out or dine out – have a meal in a restaurant

exceptionally – especially, particularly

leave – give (as in a tip)

% – percent of a number

procedure – a way to do something, a method

result (noun) – the answer or total from calculating

set amount – a fixed or determined quantity

subtotal – the total amount before another amount is added

triple (verb) – multiply by 3

check or **bill** – the paper that tells you what you owe in a restaurant

sum – total amount

sales tax – portion of the money from a sale that is paid to the government

amount tendered – the money you give a cashier

amount received – the change you get back from a cashier

tip or **gratuity** – money paid for performing a service

Dining Out

After finishing a meal in a restaurant, you receive a **bill** or **check** with the **sum** of the items that you and the people with you ordered. In some states the bill includes **sales tax.** Some states do not charge tax for restaurant food. When you pay a restaurant bill at a cash register, you get a receipt that shows the **amount tendered** (the actual amount you give to the cashier) as well as the **amount received** (change that you are owed).

When you are served at a restaurant, it is customary to leave a **tip** or a **gratuity.** There is no set amount for a tip. A common standard is 15% of the before-tax total. When service is exceptionally good, customers sometimes leave more than 15%. When service is slow or poor, customers often leave less. For large parties—say, more than six people—restaurants sometimes add a certain percentage for service.

SITUATION

Ricardo ate lunch at the Metro Diner. He had the daily chicken special for $8.95, a ginger ale for $1.35, and a piece of pie for $2.85.

EXAMPLE 1   What was the total cost of the items he ordered?

Solution   Add the costs of the three items.

$8.95 + $1.35 + $2.85 = **$13.15**

EXAMPLE 2   To estimate the cost of Ricardo's lunch, first round the cost of each item to the nearest dollar.
Then add the rounded numbers.

Solution $9 + $1 + $3 = **$13**

EXAMPLE 3   The state tax where Ricardo ate lunch is 4.5%.
What was the tax on his lunch?

Solution   Find 4.5% of $13.15.
0.045 × $13.15 = $0.59175 or **$0.59**

EXAMPLE 4   To calculate a gratuity, Ricardo used his pocket calculator
to find 15% of the pre-tax total for his food.
How much of a gratuity did Ricardo leave?

Solution   Find 15% of $13.15.

0.15 × $13.15 = $1.9725 or $1.97

Ricardo rounded this amount to $2.

EXAMPLE 5   Altogether, how much did Ricardo spend on his lunch?

Solution   Add the total cost of the items, the sales tax, and the tip.

$13.15 + $0.59 + $2.00 = **$15.74**

To solve the problems on the following pages, review:

• adding and subtracting decimals, page 230

• multiplying decimals, page 231

• finding a percent of a number, page 236

• estimation, page 228

Use the menu for Lou's Luncheonette to solve problems 1 to 8. The word *entree* refers to a main course.

Lou's Luncheonette

Burgers

Hamburger	$3.75
Cheeseburger	$4.25
Bacon Cheeseburger	$5.25

Any burger deluxe comes with French fries, coleslaw, and pickle for an additional $1.95.

Entrees

Meat Loaf	$ 7.95
Roast Turkey	$ 9.95
Broiled Chicken	$ 8.95
Roast Sirloin	$11.95

Above served with potato, vegetable, and green salad.

Beverages

Coffee	$0.75
Tea	$0.95
Iced Tea	$1.15
Iced Coffee	$1.15
Soda Pop	$1.25

Desserts

Homemade Pie	$1.85
Cake	$1.75
Ice Cream	$1.55
Custard	$1.65

Part A

Use the following information to answer problems 1 to 4.

SITUATION

Bill and Serena ate at Lou's. Bill ordered the sirloin, an iced tea, and a piece of pie. Serena ordered the turkey, an iced coffee, and a dish of ice cream.

1. What was the pre-tax total for their meals?

2. How much tax did they owe if the state charges a 6% sales tax on restaurant meals?

3. Which of the following is the best estimate for a 15% tip on the meal that Bill and Serena ate?

   (1) $2.85   (2) $3.25   (3) $3.65   (4) $4.15

4. If Bill and Serena leave a tip of $4.25, what was the total they spent for lunch including tax and tip?

Part B

Use the following information to answer problems 5 to 8.

SITUATION

Four coworkers went to eat at Lou's Luncheonette. Their order included one plain hamburger, two cheeseburgers deluxe, one plain bacon cheeseburger, two coffees, and two sodas.

5. Round each amount to the nearest dollar to estimate the total pre-tax amount.

6. Use a calculator to find the exact pre-tax total for their meals.

7. What was the average price per person for the four workers' lunches?

8. Is thirty dollars enough to cover the cost of the four lunches including a 6% sales tax and a tip? Why or why not?

Part C

Use the bill shown here to solve problems 9 to 12.

Captain Jack's			
Table 18	Guests 4	Server	Linda

2 seafood specials	$27.90
1 cola	$ 1.80
1 lemonade	$ 1.65
2 rice puddings	$ 4.70
subtotal	$36.05
4% sales tax	$ 1.44
total	$37.49

Thank You — Please Come Again

9. What was the price of one seafood special?

10. Which of the following methods for calculating a tip results in the largest amount?

   (1) Move the decimal point in the total one place to the left.

   (2) Multiply the amount of the sales tax by 4.

   (3) Move the decimal point in the subtotal one place to the left. Then divide the result by 2. Then add the two numbers.

   (4) Multiply the amount of the sales tax by 3 and add one dollar.

11. Is forty dollars enough to cover the two meals including tax and tip? Why or why not?

12. If the two diners who ate at Captain Jack's leave a tip of $5.50, what was the average amount that each one spent?

13. In a state that charges 5% sales tax on restaurant meals, which of the following is the best procedure for calculating the total cost of a meal?

   (1) Move the decimal point in the pre-tax total one place to the left.

   (2) Triple the sales tax to calculate the tip.

   (3) Leave a tip that is equal to the amount of the sales tax.

   (4) Add $2 to the amount of the sales tax to calculate the amount of the tip.

Post-Lesson Vocabulary Reinforcement

Choose the appropriate vocabulary word from the list on the right to complete the sentences.

1. When served in a restaurant, it is customary to leave a _____.

2. Leaving only 5% is not an _____ tip.

3. This total does not include the tax; it's a _____.

4. When service is _____ good, customers sometimes tip more than 15%.

5. I'll _____ a big tip because the service was very good.

6. First _____ the cost of each item to the nearest dollar. Then add the rounded numbers.

7. _____, how much did Ricardo spend on his lunch?

8. Thirty dollars is enough to _____ the cost of lunch.

9. Restaurants sometimes add a certain _____ for service.

10. Move the decimal point in the subtotal one place to the left. Then divide the _____ by 2.

11. Before you leave a restaurant, remember to pay the bill _____ leave a tip.

12. The amount you owe a restaurant is listed on the _____.

adequate

all together

as well as

before-tax total

check or bill

cover

exceptionally

leave

percentage

result

round

tip or gratuity

Match each definition with the correct term.

_____ 13. to multiply by three

_____ 14. a way of doing something

_____ 15. the mean or midpoint

_____ 16. an acceptable amount

_____ 17. a fixed quantity

_____ 18. use numbers to figure an amount

_____ 19. the amount before tax and tip are added

_____ 20. to dine out

_____ 21. the money you give a cashier

_____ 22. money on a bill that goes to the government

_____ 23. the total of a bill or check

_____ 24. the money you get back as change

a. set amount

b. common standard

c. calculate

d. average

e. subtotal

f. procedure

g. triple

h. eat out

i. amount tendered

j. amount received

k. sum

l. sales tax

Language Builder

ACTIVITY A Read the first sentence in the lesson. The verb *receive* is in the **present tense.** The verb *ordered* is in the **past tense.** Read through the lesson and add verbs to the list below.

Present Tense Verbs	**Past Tense Verbs**
receive	ordered

ACTIVITY B Notice that a present tense and a past tense verb can occur in the same sentence: "After finishing a meal in a restaurant, you *receive* a bill or check with the sum of the items that you and the people with you *ordered.*" That's because the verb *receive* refers to an action that usually occurs (present tense), while the verb *ordered* refers to an action that occurred in the past (past tense).

Practice using the present tense and the past tense in the same sentence. On a separate sheet, write the correct form of the infinitive verbs in parentheses.

1. Here's the bill. You only (have) to pay for the food that you (eat).

2. When you go to a restaurant, you (get) a receipt that shows the amount (tender).

3. The state tax where Ricardo (have) dinner yesterday (be) 4.5%.

4. The standard tip (be) 15%, but the service was poor, so Ricardo (leave) only 10%.

5. I (get) a total of $1.97 that Lou (round) off to $2.

ACTIVITY C The command form of the verb is the **infinitive form** of the verb. Read the sentences about people who did the wrong thing. Tell them to do the right thing. Use the words in parentheses and the command form of the verb to write a new sentence on a separate sheet.

6. Ricardo moved the decimal point one place to the right. (to the left)
   <u>Move the decimal point one place to the left.</u>

7. Jack subtracted the tax from the bill. (add . . . to the bill)

8. Jack solves the problems in one exercise. (all the exercises)

9. Bill and Serena paid part of the bill. (all of the bill)

10. Lou doubled the sales tax to calculate the tip. (triple)

Now practice the meanings of the command forms by writing your own sentences using each form from the list you made. Then compare your work with a partner and take turns acting out each other's commands as you say them.

LESSON 6: TRAVEL

Pre-Lesson Vocabulary Practice

The words on the right appear in Lesson 6. Read the words and the definitions below. Next find them in the lesson. Carefully read the lesson material to better understand their meanings. Then match each term to its definition.

_____ 1. clock times that state when a train, bus, etc., reaches its destination

_____ 2. schedules that show the times when things happen

_____ 3. clock times that state when a train, bus, etc., leaves

_____ 4. "kinds" or "types" of something

_____ 5. an amount of money one has to pay for a transportation service

_____ 6. trips that are arranged according to specific times

_____ 7. the mechanism that displays the taxi fare

_____ 8. parts of something

_____ 9. an amount of money charged

a. **arrival times**

b. **departure times**

c. **fare**

d. **fee**

e. **meter**

f. **modes**

g. **scheduled trips**

h. **segments**

i. **timetables**

Work with a partner. Use each of the terms listed below in a sentence of your own.

blocked traffic or **stuck in traffic** – vehicles that can't move

common practice – something people usually do

delay or **waiting time** – the cost charged by a taxi when it's not moving

evenly – equally, the same amount per person

express trains (E) – trains that go from one place to another without stopping

initial fee – the first amount of money charged

local trains (L) – trains that travel only within a specific area or region

monthly pass – travel fare for an entire month

off-peak – a time when a lot of people do not travel

on-board fare – the price charged for a ticket on the train

one-way (OW) – in only one direction

peak – a ticket for travel during a time when a lot of people travel; peak ticket

round-trip (RT) – to go both "to" and "from" a particular place

running on schedule – keeping the schedule; staying according to the schedule

Travel

Public transportation includes travel on buses, taxis, subways, trains, ferries, and airplanes. The **fare** that a customer pays to ride some **modes** of public transportation, such as ferries and airplanes, is set by the companies that own and operate the equipment. For taxis, local agencies control the licensing of the drivers and the fares that are charged.

The cost of a ride in a taxi varies from community to community. Often there is an initial **fee** as well as an additional fee for each fraction of a mile the customer travels. In the United States it is customary to give a taxi driver a tip of 10% to 15% of the amount on the taxi **meter.** (You will learn more about the costs of the most popular form of private transportation, the automobile, later in this book.)

EXAMPLE 1   In Boston, the cost of a taxi ride is $1.80 for the first $\frac{1}{8}$ mile and $0.30 for each additional $\frac{1}{8}$ mile. What is the cost of a $2\frac{1}{2}$-mile taxi ride in Boston including a 15% tip?

Solution   First find the number of $\frac{1}{8}$-mile segments in $2\frac{1}{2}$ miles.

$2\frac{1}{2} \div \frac{1}{8} = \frac{5}{2} \times \frac{8}{1} = \frac{40}{2} = 20$ segments

The first $\frac{1}{8}$ mile costs $1.80.

The next 19 segments cost 19 × $0.30 = $5.70.

The total fare is $1.80 + $5.70 = $7.50.

A 15% tip is 0.15 × $7.50 = $1.125 or $1.13.

The fare including a 15% tip is $7.50 + $1.13 = **$8.63.**

It is a common practice to round a number like $8.63 to a more convenient number in our money system, such as $8.75 or even $9.00.

Train, airplane, and bus companies publish **timetables** that show **departure times** and **arrival times** for their **scheduled trips.** It is sometimes difficult to determine how much time a trip requires. Travel time is simply the arrival time minus the departure time.

EXAMPLE 2   If Ruth gets on the bus at her corner at 8:05 in the morning and gets off near her office at 8:30 in the morning, what is her travel time?

Solution   To find Ruth's travel time, subtract.

8:30 − 8:05 = **25 minutes**

To calculate travel time for a trip, remember that an hour has 60 minutes and that a day has 24 hours. We usually divide the 24 hours in a day into two twelve-hour periods. To calculate travel time, it is sometimes convenient to use a 24-hour clock.

On a 24-hour clock, any time before noon remains the same, but time after noon and before midnight is 12 plus the time.

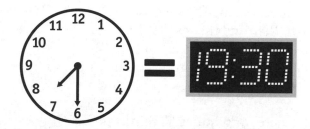

On a 24-hour clock, 7:30 P.M. is: 12:00 + 7:30 = 19:30.

EXAMPLE 3   Amtrak, the national train company, has a train that leaves Rochester, New York, at 11:47 A.M. and arrives in New York City at 6:40 P.M. If the train is running on schedule, what is the total travel time from Rochester to New York City?

ROCHESTER, NY TO NEW YORK CITY		
MONDAY TO FRIDAY, EXCEPT HOLIDAYS		
Leaves	Rochester, NY	11:47 A.M.
Arrives	New York City, NY	6:40 P.M.

Solution   First find the arrival time on a 24-hour clock.

6:40 P.M. = 12:00 + 6:40 = 18:40

Subtract the departure time from the arrival time:

$$18:40$$
$$-11:47$$

To subtract 47 from 40, borrow 1 of the 18 hours and change it to 60 minutes. Then add the 60 minutes to 40 minutes.

Arrival time = 18 hr 40 min =	17 hr	100 min
Departure time =	− 11 hr	47 min
Total travel time =	**6 hr**	**53 min**

There is another way to solve the last problem. Find the travel time before noon and the travel time after noon. Then add the results. The departure time is 11:47 A.M. or 13 minutes before noon. The arrival time is 6:40 P.M. or 6 hours and 40 minutes after noon. The total travel time is:

6 hours 40 minutes + 13 minutes = **6 hours 53 minutes**

To solve the problems in the next exercise, review:

- dividing fractions, page 235
- multiplying decimals, page 231
- subtracting measurements, page 239

EXERCISE 6

Part A

Use a calculator to solve the problems. Use the following information to answer problems 1 to 3.

SITUATION

> A taxi ride in Philadelphia costs $1.80 from the time the meter is turned on for the first $\frac{1}{6}$ mile and $0.30 per $\frac{1}{6}$ mile thereafter. The waiting time rate is $0.20 per minute.

1. What is the basic cost of a $1\frac{1}{2}$-mile taxi ride in Philadelphia?

2. What is the cost of a $2\frac{3}{4}$-mile taxi ride that includes six minutes of stopping time for heavy traffic and a 15% tip?

3. Find the cost, including a 15% tip, of a taxi ride that is only $\frac{1}{2}$ mile long, but includes a three-minute delay for street construction.

Part B

Use the following information for problems 4 and 5.

SITUATION

> In Chicago, the fare for a taxi ride starts at $1.90. The cost is then $1.60 per mile plus a $2.00 charge for every 6 minutes of waiting time. There is a $0.50 charge for each additional passenger over 12 and under 65.

4. Find the basic cost for a $2\frac{1}{2}$-mile taxi ride in Chicago for two people if there is no waiting time.

5. Find the cost including a 15% tip for a $6\frac{1}{4}$-mile taxi ride for a single passenger in Chicago if the ride is stopped for a total of 12 minutes because of a traffic accident.

Part C

The table below and the table on the next page show the weekday timetable and fares for trains from White Plains to New York City. *E* is for express trains, and *L* is for local trains. *X* means that the train does not stop at 125th Street, and *OW* means one-way. *Peak* refers to trains that leave White Plains from 5:03 A.M. until 9:03 A.M. Use the tables to answer problems 6 to 11.

WHITE PLAINS TO NEW YORK					
MONDAY TO FRIDAY, EXCEPT HOLIDAYS					
LEAVE	ARRIVE	LEAVE	ARRIVE	LEAVE	ARRIVE
White Plains	New York	White Plains	New York	White Plains	New York
AM	**AM**	**AM**	**AM**	**PM**	**PM**
12:06 L	12:55	X 9:00 E	9:33	**4:06 L**	**4:59**
5:03 L	5:48	9:03 E	9:42	**4:25 E**	**5:04**
5:31 E	6:02	9:26 E	10:03	**4:31 L**	**5:21**
5:35 L	6:20	9:29 L	10:18	**4:58 E**	**5:35**
5:59 E	6:30	9:54 E	10:33	**5:03 L**	**5:56**
6:04 L	6:49	9:58 E	10:36	**5:32 E**	**6:10**
6:19 L	7:03	10:08 L	11:01	**5:38 L**	**6:29**
6:30 E	7:05	10:30	11:14	**5:58 E**	**6:36**
6:35 E	7:09	10:58 E	11:34	**6:08 L**	**7:00**
6:40 E	7:13	11:08 L	11:59	**6:36**	**7:21**
6:43	7:26	11:30	**12:14**	**6:51 E**	**7:27**
6:51 E	7:29	11:58 E	**12:34**	**7:06 L**	**7:55**
7:00 E	7:33	**12:08 L**	**1:01**	**7:58 E**	**8:32**
7:08	7:50	**12:30**	**1:14**	**8:06 L**	**8:57**
X 7:18 E	7:52	**12:58 E**	**1:35**	**8:58 E**	**9:32**
7:22	7:56	**1:08 L**	**1:59**	**9:06 L**	**9:55**
7:37 E	8:14	**1:30**	**2:14**	**9:58 E**	**10:32**
7:40 E	8:18	**1:58 E**	**2:34**	**10:06 L**	**10:57**
X 8:01 E	8:38	**2:08 L**	**3:01**	**10:58 E**	**11:32**
8:05 E	8:42	**2:30 L**	**3:22**	**11:06 L**	**11:55**
8:13 E	8:51	**2:58 E**	**3:34**	12:06 L	12:55
8:28 E	9:05	**3:08 L**	**3:59**	--:--	--:--
8:31 E	9:07	**3:33 L**	**4:23**	--:--	--:--
X 8:40 E	9:19	**3:58 E**	**4:34**	--:--	--:--
AM	**AM**	**PM**	**PM**	**AM**	**AM**

SAMPLE FARES TO GRAND CENTRAL TERMINAL & HARLEM–125TH STREET

Ticket Type:	Web Ticket	Station	On Board Train
OW Peak:	$7.60	$8.00	$11.00
10 Trip Peak:	$76.00	$80.00	N/A
OW Off-Peak:	$5.70	$6.00	$9.00
10 Trip Off-Peak:	$48.45	$51.00	N/A
Monthly:	$171.50	$175.00	N/A
Weekly:	$53.20	$56.00	N/A

6. Find the time in minutes for each of the following train trips from White Plains to New York:

   a. the local train that leaves White Plains at 6:19 A.M.

   b. the express train that leaves White Plains at 8:40 A.M.

   c. the local train that leaves White Plains at 5:38 P.M.

   d. the express train that leaves White Plains at 8:58 P.M.

7. What is the cost per ride of a 10-trip peak ticket that is bought on the Web?

8. What is the cost per ride of a 10-trip peak ticket that is bought at the station?

9. Find the cost per ride of a 10-trip off-peak ticket that is purchased on the Web.

10. Find the cost per ride of a 10-trip off-peak ticket that is purchased at the station.

11. In one month with no holidays, Rosa calculated that she will make 22 round-trip rides between White Plains and New York or a total of 44 rides. To the nearest cent, what will be the cost per ride if she buys a monthly pass for the trips between White Plains and New York on the Web?

Part D

The charges for a taxi ride in New York City are shown below. Use the information to answer problems 12 to 14.

```
┌─────────────────────────────────────────────┐
│              TAXI FARE                        │
│  Initial fare                    $2.00        │
│  Each 1/5 mile (4 blocks)        $0.30        │
│  Each 1 minute idle              $0.20        │
│  Night surcharge                 $0.50        │
│  (after 8 P.M. until 6 A.M.)                  │
│  Additional riders               FREE         │
└─────────────────────────────────────────────┘
```

12. What is the cost, including a 15% tip, of a 5-mile taxi ride in New York City at lunch time if there are no delays?

13. Find the cost, including a 15% tip, of a $1\frac{1}{2}$-mile taxi ride in New York City at 8:30 P.M. if there are 8 minutes of waiting time because of blocked traffic.

14. Three businessmen share a taxi in New York City at 9 o'clock at night. Their ride is 10 miles long, and they get stuck in traffic for fifteen minutes. Approximately how much do they each owe if they divide the fare and a 15% tip evenly?

Part E

The timetable below shows the times of a train that travels between Chicago and New Orleans. To read the times between Chicago and New Orleans, read down the column at the left. To read the times between New Orleans and Chicago, read up the column at the right. Use the timetable to answer problems 15 to 23.

Read Down	Mile	▼	Chicago • Memphis • New Orleans	Symbol	▲	Read Up
59			◄ Train Number ►			**58**
Daily			◄ Days of Operation ►			**Daily**
R 🛏 ✕ ☕			◄ On Board Service ►			R 🛏 ✕ ☕
🧳 **8 00P**	0	Dp	**Chicago, IL**–Union Sta. (CT) 🚌 Madison—see page 42	♿ ☀	Ar	🧳 **9 00A**
19 **8 54P**	25		**Homewood, IL** (METRA/IC Line)	⊘		19 **7 44A**
✕ **9 23P**	57		**Kankakee, IL**	● ♿		✕ **7 13A**
🧳 **10 34P**	129		**Champaign-Urbana, IL**	⊘ ♿		**6 10A**
✕ **11 13P**	174		**Mattoon, IL** (Charleston)	●		✕ **5 23A**
✕ **11 37P**	201		**Effingham, IL**	● ♿		✕ **4 57A**
✕ **12 25P**	254	▼	**Centralia, IL**	● ♿		✕ **4 10A**
🧳 **1 21A**	310	Ar	**Carbondale, IL**	♿	Dp	🧳 **3 16A**
🧳 **1 26A**		Dp	🚌 St. Louis, Kansas City—see below		Ar	🧳 **3 11A**
✕ **3 14A**	407		**Fulton, KY**	● ♿	▲	✕ **1 04A**
✕ **3 56A**	442	▼	**Newbern-Dyersburg, TN**	● ♿		✕ **12 22A**
🧳 **6 27A**	520	Ar	**Memphis, TN**	♿ ☀	Dp	🧳 **10 40P**
🧳 **6 50A**		Dp			Ar	🧳 **10 00P**
9 00A	644		**Greenwood, MS**	● ♿	▲	**7 37P**
✕ **9 51A**	697		**Yazoo City, MS**	● ♿		✕ **6 42P**
🧳 **11 20A**	741		**Jackson, MS**	♿		**5 44P**
✕ **11 56A**	777		**Hazlehurst, MS**	● ♿		✕ **4 33P**
✕ **12 18P**	797		**Brookhaven, MS**	● ♿		✕ **4 13P**
✕ **12 45P**	821		**McComb, MS**	● ♿		✕ **3 48P**
1 43P	873	▼	**Hammond, LA**	♿		**2 58P**
🧳 **3 40P**	926	Ar	**New Orleans, LA** (CT) 🚌 Baton Rouge, Mobile—see below	♿ ☀	Dp	🧳 **1 55P**

15. What is the total time for the train trip from Chicago to New Orleans?

16. What is the total time for the train trip from New Orleans to Chicago?

17. On the trip from Chicago to New Orleans, how long is the layover—the time between arriving and departing—in Memphis?

18. A one-way coach ticket between Chicago and New Orleans costs $142. To the nearest tenth of a cent, what is the cost per mile of the train trip?

19. How long is the trip from Mattoon, IL, to Fulton, KY?

20. How long does it take to travel from McComb, MS, to Memphis, TN?

21. When traveling from Effingham, IL, to Newbern-Dyersburg, TN, how long is the layover in Carbondale, IL?

22. How many miles is it from Centralia, IL, to Hammond, LA?

23. What is the distance from Jackson, MS, to Homewood, IL?

Post-Lesson Vocabulary Reinforcement

Match the terms on the right with the correct blanks.

1. _____

2. _____

3. _____

4. _____

5. _____

departure time

train schedule

departure city

arrival city

arrival time

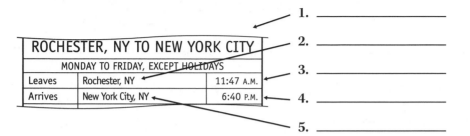

ROCHESTER, NY TO NEW YORK CITY		
MONDAY TO FRIDAY, EXCEPT HOLIDAYS		
Leaves	Rochester, NY	11:47 A.M.
Arrives	New York City, NY	6:40 P.M.

WHITE PLAINS TO NEW YORK					
MONDAY TO FRIDAY, EXCEPT HOLIDAYS					
LEAVE	ARRIVE	LEAVE	ARRIVE	LEAVE	ARRIVE
White Plains	New York	White Plains	New York	White Plains	New York
AM	AM	AM	AM	PM	PM
12:06 L	12:55	X 9:00 E	9:33	4:06 L	4:59
5:03 L	5:48	9:03 E	9:42	4:25 E	5:04
5:31 E	6:02	9:26 E	10:03	4:31 L	5:21
5:35 L	6:20	9:29 L	10:18	4:58 E	5:35
5:59 E	6:30	9:54 E	10:33	5:03 L	5:56
6:04 L	6:49	9:58 E	10:36	5:32 E	6:10
6:19 L	7:03	10:08 L	11:01	5:38 L	6:29
6:30 E	7:05	10:30	11:14	5:58 E	6:36
6:35 E	7:09	10:58 E	11:34	6:06 L	7:00
6:40 E	7:13	11:08 L	11:59	6:36	7:21
6:43	7:26	11:30	12:14	6:51 E	7:27
6:51 E	7:29	11:58 E	12:34	7:06 L	7:55
7:00 E	7:33	12:08 L	1:01	7:58 E	8:32
7:08	7:50	12:30	1:14	8:06 L	8:57
X 7:18 E	7:52	12:58 E	1:35	8:58 E	9:32
7:22	7:56	1:08 L	1:59	9:06 L	9:55
7:37 E	8:14	1:30	2:14	9:58 E	10:32
7:40 E	8:18	1:58 E	2:34	10:06 L	10:57
X 8:01 E	8:38	2:08 L	3:01	10:58 E	11:32
8:05 E	8:42	2:30 L	3:22	11:06 L	11:55
8:13 E	8:51	2:58 E	3:34	12:06 L	12:55
8:28 E	9:05	3:08 L	3:59	--:--	--:--
8:31 E	9:07	3:33 L	4:23	--:--	--:--
X 8:40 E	9:19	3:58 E	4:34	--:--	--:--
AM	AM	PM	PM	AM	AM

6. _____

7. _____

a. **local train**

b. **express train**

c. **one-way off-peak fare**

d. **one-way peak fare**

e. **on-board fare**

SAMPLE FARES TO GRAND CENTRAL TERMINAL & HARLEM–125TH STREET

Ticket Type:	Web Ticket	Station	On Board Train
OW Peak:	$7.60	$8.00	$11.00
10 Trip Peak:	$76.00	$80.00	N/A
OW Off-Peak:	$5.70	$6.00	$9.00
10 Trip Off-Peak:	$48.45	$51.00	N/A
Monthly:	$171.50	$175.00	N/A
Weekly:	$53.20	$56.00	N/A

8. _____

9. _____

10. _____

Choose terms from the list on the right to complete the sentences that follow.

11. _____ includes travel on buses, taxis, subways, trains, ferries, and airplanes.

12. For one-fifteen P.M. a(n) _____ clock shows 1:15.

13. For one-fifteen P.M. a(n) _____ clock may show 13:15.

14. The time between arriving on one train and departing on another is called _____.

15. An example of _____ travel is travel between 2 P.M. and 3 P.M.

analog clock
(24-hour clock)

digital clock

public transportation

off-peak

a layover

Language Builder

ACTIVITY A Remember the **passive form** of the verb from Lesson 3? This lesson has some good examples for review. Remember that what makes a verb passive is the **verb to be** (is / are) + the **past participle** of the verb. <u>Underline</u> the form of the **verb to be** + the **past participle** in each of the following sentences adapted from this lesson.

1. From the time the meter is turned on, a taxi ride in Philadelphia costs $1.80.

2. If a taxi ride is stopped for an accident, the waiting-time rate takes effect.

3. How much is a train ticket that is purchased on the Web?

4. Find the cost of two tickets that are bought at the station.

Now underline the basic subject of each passive verb twice in the above sentences. Here's an example:

<u>Licensing</u> is <u>controlled</u> by local agencies.

ACTIVITY B Complete the following sentences by making the verb in parentheses passive.

5. Those fares (control) _____ by the companies that own the taxis.

6. A fare (charge) _____ even if the taxi (delay) _____ for street construction.

7. Many taxis (stop) _____ because of heavy traffic.

ACTIVITY C The word *if* can begin a type of sentence that is located inside another sentence. An **if clause** is the name for this kind of sentence. It shows that a condition exists—something depends on, or is determined by—something else. Here's an example:

Find the basic cost of the taxi ride <u>if there is no waiting time.</u>

Put the words for the if clauses in the correct order and write them in the blanks.

8. What is the cost, including a 15% tip, of a 5-mile taxi ride in New York City at lunch time (there delays are if no) **_if there are no delays?_**

9. Find the basic cost for a $2\frac{1}{2}$-mile taxi ride in Chicago (is if waiting no time there) _____ .

10. Find the basic cost of a $1\frac{1}{2}$-mile taxi ride in New York City at 8:30 P.M. (there eight minutes are if of waiting time) _____ because of blocked traffic.

11. Approximately how much do they each owe (the fare and if a 15% tip evenly divide they) _____ ?

Pre-Lesson Vocabulary Practice

Read the terms on the right and find them in the lesson. Then carefully read the lesson material to learn the meaning of each term. Next match the terms with their definitions below.

_____ **1.** repairing, fixing, or renewing a house

_____ **2.** people whose houses belong to them

_____ **3.** people who pay money on a regular basis to occupy a piece of property

_____ **4.** work done on a piece of property by oneself

_____ **5.** work done to make a house better

_____ **6.** to increase the worth of a piece of property

_____ **7.** to enclose or contain something

_____ **8.** an amount of space or area specified

_____ **9.** almost, nearly, more or less

_____ **10.** to construct

_____ **11.** the cost of sending something from one place to another

a. approximately

b. coverage

c. delivery charge

d. do-it-yourself projects

e. home improvement projects

f. home owners

g. home renovation

h. renters

i. to build

j. to raise the value of a property

k. to surround

Work with a partner to use the boldfaced words below in a sentence.

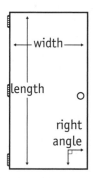

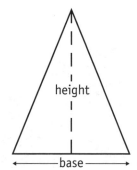

A **rectangle** is a four-sided figure in which the opposite sides are **parallel,** and the sides meet to form **right angles.** Most doors are rectangles. The **length** is usually the longer of the two **dimensions** and the **width** is the shorter.

Triangles and rectangles make up most **interior** and **exterior** walls. A triangle is a flat figure with three sides. The **base** is the triangle's greatest **horizontal** measure. The **height** is the greatest **vertical** measure.

Perimeter is a measure of the distance around a flat figure. Perimeter is measured in **linear** units such as feet, inches, yards, and meters.

Note:
Linear Units

a foot = 12 inches

a yard = 3 feet

a meter = 39.37 inches

Home Renovation

Many home owners and renters improve their surroundings with do-it-yourself projects. From the relatively simple task of painting the walls and ceiling of a bedroom to the more complicated job of adding a new room to a house, home improvement projects are good ways to raise the value of a property.

To estimate the quantity of materials that are needed for a project and to get an idea of the cost, a consumer needs to use some basic principles from geometry.

Perimeter is a measure of the distance around a flat figure. Perimeter is measured in **linear** units such as feet, inches, yards, and meters. The formula for the perimeter of a rectangle is

$$P = 2l + 2w$$

where P is the perimeter, l is the length, and w is the width. A **rectangle** is a four-sided figure in which the opposite sides are **parallel,** and the sides meet to form **right angles.** Most doors, windows, floors, and ceilings are rectangles. The **length** is the longer of the two **dimensions** of a rectangle. The **width** is the shorter dimension.

EXAMPLE 1 What is the perimeter of a rectangular vegetable garden that is 15 feet long and 9 feet wide?

9 ft

15 ft

Solution Replace *l* with 15 ft and *w* with 9 ft in the formula for the perimeter of a rectangle.

$$P = 2(15) + 2(9) = 30 + 18 = \textbf{48 feet}$$

The cost formula is $c = nr$ where c is the cost, n is the number of units, and r is the rate, or cost, for one unit. To estimate the cost of most home renovation projects, use the cost formula.

EXAMPLE 2 A garden supply store sells wood fencing at $2.49 per linear foot. What is the total cost of the fencing required to surround the vegetable garden in the last example?

Solution $c = nr$ = 48 ft × $2.49 per foot = **$119.52**

The message "Coverage: approximately 50 sq ft depending on method of application and color" appears on a one-pint can of paint. The abbreviation *sq ft* means **square feet,** which is a unit of measure for **area.** Area is a measure of flat surface. Area is measured in units such as square inches, square feet, square yards, and square meters. Most **interior** and **exterior** walls are **rectangles** or a combination of rectangles and **triangles.** A triangle is a flat figure with three sides.

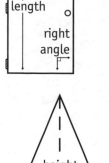

For a rectangle, the formula for the area is $A = lw$ where A represents the area, l represents the length, and w represents the width.

For a triangle, the formula for the area is $A = \frac{1}{2}bh$ where A is the area, b is the base, and h is the height. The **base** is the triangle's greatest **horizontal** measure. The **height** is the greatest **vertical** measure.

EXAMPLE 3 The illustration shows an end wall of a garden shed. What is the area of the wall including the door?

Solution The wall of the garden shed is composed of two geometric shapes: a triangle (labeled A) and a rectangle (labeled B).

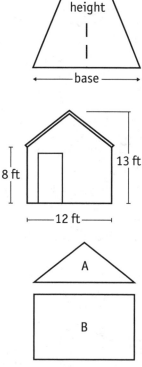

The triangle has a base of 12 feet. The height of the triangle is the total height of the shed (13 ft) minus the height of the rectangle (8 ft). The height of the triangle is $13 - 8 = 5$ feet.

The area of the triangle is $A = \frac{1}{2} \times 12 \times 5 = 30$ sq ft.

The length of the rectangle is 12 feet, and the width is 8 feet.

The area of the rectangle is $A = 12 \times 8 = 96$ sq ft.

The total area of the wall is $30 + 96 = $ **126 sq ft.**

EXAMPLE 4   If one pint of paint can cover 50 square feet, how many pints of paint are required to cover the end wall and door of the shed in Example 3?

Solution   Divide the area by 50, the number of square feet that one pint will cover.

126 ÷ 50 = 2.52 pints

The painter should buy **3 pints** of paint.

EXAMPLE 5   If one pint of paint costs $3.98, what is the cost of the paint needed to cover the end wall of the shed?

Solution   **3 × $3.98 = $11.94**

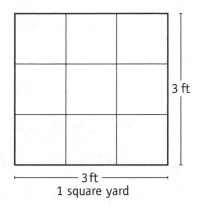
3 ft
3 ft
1 square yard

Square feet is the most common unit of measure for the area of building materials, but some products are traditionally measured in other units. Carpet, for example, is usually priced by the **square yard.** A square yard measures 3 feet on each side. One square yard has an area of 3 × 3 = 9 square feet.

EXAMPLE 6   How many square yards of carpeting are required to cover the floor of a bedroom that is 12 feet wide and 14 feet long?

Solution   The area of the floor is 14 × 12 = 168 square feet. The area is 168 ÷ 9 = 18.666... or about **18.7 sq yd.**

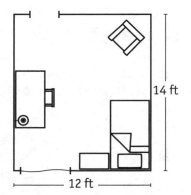
14 ft
12 ft

EXAMPLE 7   At the price of $24.95 a square yard, what is the cost of the carpet for the floor of the room in the last example?

Solution   *c* = *nr* = 18.7 square yards × $24.95 per square yard = $466.565 or **$466.57**

To solve the problems in the next exercise, review:

- finding the perimeter of a rectangle, page 241
- finding the area of a rectangle, page 241
- finding the area of a triangle, page 242
- multiplying decimals, page 231
- multiplying fractions, page 235

Part A

Use a calculator to solve the problems.

For problems 1 to 6 tell whether you need to find the perimeter or the area to estimate the quantity of materials that are required for each home improvement project.

1. the amount of wood trim to go around a new kitchen window

2. the number of ceramic tiles to go on a bathroom wall

3. the amount of new hardwood flooring for a living room

4. the number of bricks needed to make a border for a flower garden

5. the amount of cedar siding to cover the walls of a new garage

6. the amount of stain to apply to the boards on a deck

Part B

The illustration shows the front wall and one side wall of Carl's garage. Use the illustration to answer problems 7 to 11.

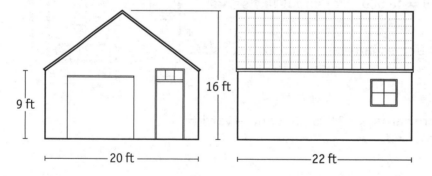

7. What is the area of the front wall of the garage including the two doors?

8. What is the area of the side wall of the garage including the window?

9. What is the total surface area of the garage walls not including the roof?

10. One gallon of the paint that Carl plans to use on his garage covers about 350 square feet. How many gallons of paint should Carl buy to give the doors, siding, and trim one coat of paint?

11. The paint that Carl plans to use on his garage sells for $22.95 per gallon. What is the total cost of enough paint to give the garage one coat?

Part C

Use the following information to answer problems 12 and 13.

SITUATION

Maureen wants to build simple wooden frames around two posters in her son's room. Each poster is a rectangle that is $4\frac{1}{2}$ feet tall and $2\frac{1}{2}$ feet wide. To build the frames, Maureen plans to use wood molding that costs $2.89 per linear foot.

12. Approximately how many feet of wood molding does Maureen need to build the frames?

13. Maureen decided to buy 2 more feet of molding than she calculated that she needs. What is the total cost of the molding, including a sales tax of 6.5%?

14. After she finished building the frames for the posters, Maureen decided to put wall-to-wall carpet on the floor of her son's room. Her son's room is $10\frac{1}{2}$ feet wide and 14 feet long. How many square yards of carpet does she need to cover the floor of the room?

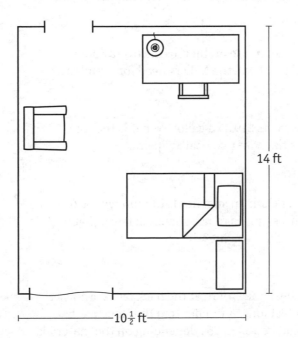

14 ft

$10\frac{1}{2}$ ft

15. The carpet that Maureen wants to buy costs $32.95 per square yard. What is the cost of the carpet including a $15 delivery charge?

Part D

The illustration shows the dimensions of the barn on the Miller family's farm. Use the illustration to answer problems 16 to 18.

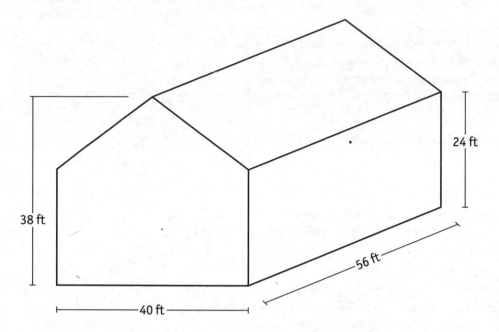

16. What is the total surface area of the walls of the barn? (Do not include the roof.)

17. A gallon of the paint that the Millers plan to use on their barn covers about 400 square feet. How many gallons of paint will the Millers need for their barn?

18. The paint that the Millers plan to buy costs $19.85 a gallon. What is the total cost of the paint if they decide to give the barn two coats of paint?

19. Felidia wants to build a dog run for her German shepherd. The run will be 65 feet long and 5 feet wide. To build the run, she plans to use chainlink fence that costs $0.89 per linear foot. What is the cost of the fencing?

20. The Nocera family wants to build a new family room at the back of their house. The room will be 18 feet wide and 24 feet long. A builder told the Noceras that the square-foot cost for the room could be $32 to $54 depending on the materials that they choose. Find the difference between the builder's lowest and highest estimates for the total cost of the project.

Post-Lesson Vocabulary Reinforcement

Choose the appropriate term from the list at the right to complete the sentences.

1. A _____ of the paint that the Millers plan to use covers about 400 square feet.

2. The message "_____: approximately 50 sq ft depending on method of application and color" appears on a one-pint can of paint.

3. The title of this lesson is _____.

4. _____ are examples of home improvement projects.

5. Many _____ and _____ improve their surroundings with do-it-yourself projects.

6. Maureen wants _____ simple wooden frames around two posters.

7. Home improvement projects are good ways _____.

8. The perimeter of the vegetable garden is represented by the fencing required _____ it.

9. In Part D, you were asked to _____ between the builder's lowest and highest estimates.

10. The cost of the carpet includes a $15 _____.

11. Please estimate _____ how many feet of wood molding Maureen needs.

approximately

coverage

delivery charge

do-it-yourself projects

find the difference

gallon

home owners

home renovation

renters

to build

to raise the value of a property

to surround

Complete each sentence with the correct word.

width	length	rectangle	triangle
foot	meter	yard	

12. This shape is a **rectangle**.

13. This is the _____.

14. This is the _____.

15. This shape is a _____.

16. This is the _____ measure.

17. This is the _____ measure.

18. A _____ equals 12 inches.

19. A _____ equals 39.37 inches.

20. A _____ equals 3 feet.

height

base

Language Builder

W/H information questions begin with a <u>W</u> question word such as *What* . . . ? or the <u>H</u> question word *How* . . . ? Examples from this lesson are: *What <u>is</u> the area of the wall including the door?* and *How many square yards of carpeting <u>are</u> required?* Notice that the underlined words are a form of the verb **to be.** Other verbs are also used that follow a different pattern. For example: *How many square yards of carpet <u>does</u> she <u>need</u> to cover the floor of the room?*

In the blanks write the appropriate question word—**What** or **How**:

1. _____ is the total cost of the fencing required to surround the garden?

2. _____ much does John make in overtime wages in a week?

3. _____ many square yards of carpeting are required to cover the floor?

4. _____ are the measurements of that wall?

ACTIVITY B Now choose verb forms from the box to fill in the blanks of the following W/H information questions from this lesson.

is	does	will

5. Approximately how many feet of wood molding _____ Maureen need to build the frames?

6. What _____ the area of the front wall of the garage including the two doors?

7. How many gallons of paint _____ the Millers need for their barn?

ACTIVITY C Here is a review of the **infinitive of purpose.** If you can substitute the words "in order to" for the infinitive (*to + verb*), then it is an infinitive of purpose. The two infinitives of purpose in the following sentence are underlined.

<u>To estimate</u> the quantity of materials that are needed for a project and <u>to get</u> an idea of the cost, a consumer needs to use some basic principles from geometry.

In the following sentences, underline the infinitives of purpose.

8. To raise the value of a property, many owners use do-it-yourself projects.

9. What is the total cost of the fencing required to surround the vegetable garden?

10. How much paint is needed to cover the end wall of the shed?

Pre-Lesson Vocabulary Practice

On the right are some terms from this lesson that are not defined in the text. Find the terms in the lesson, and read the surrounding sentences to try to better understand their meanings. Finally, match the terms on the right with their definitions below.

_____ 1. someone, such as a stockbroker, who gives advice or counsel

_____ 2. is complete after a fixed amount of time

_____ 3. places where stocks are bought and sold

_____ 4. equitable claims, stocks

_____ 5. a person who buys and sells stocks for investors

_____ 6. persons who invest

_____ 7. an amount gained or earned after any deductions or charges

_____ 8. a shared holding or ownership

_____ 9. to put money in a transaction in order to get a financial return

_____ 10. money that a person makes or earns

_____ 11. three-month segments of a calendar year

_____ 12. putting money in a transaction, the transaction itself

a. **financial return**

b. **to invest**

c. **investing / investment**

d. **investors**

e. **an equitable claim**

f. **shares**

g. **stockbroker / broker**

h. **an advisor**

i. **exchanges**

j. **quarters**

k. **matures**

l. **net profit**

Carefully read the line from the newspaper stock page and the information about it on pages 82–83. Then write the correct letter in each blank.

| 52-Week | | | | Yld | | Sales | | | | |
High	Low	Stock	Div.	%	P/E	100s	High	Low	Last	Chg.
27.11	15.54	MLX	.30	1.2	26	5254	25.85	24.90	25.84	+0.26

13. _____   14._____   15._____ 16.___ 17.____   18._____ 19._____ 20._____

a. amount a company expects to pay for each share

b. difference between the last price of the stock on the previous day and the last price on the current trading day

c. Marlex

d. highest value of the stock in the past year

e. yield—the annual dividend divided by the closing price

f. the price-to-earnings ratio

g. the lowest price of a share of Marlex for the day

h. last trading price for the day— closing price

Investing

A consumer **invests** money in order to get a financial **return.** There are many books, magazines, and newspaper articles that give readers tips on how to invest their money. But the basic mathematics of investing is simple. The goal for an investor is to buy for less and sell for more. The specialized vocabulary used by **investors** and their **advisors** can be overwhelming at first. In this lesson you will learn some of the basic terms about investing.

Four popular forms of investment are stocks, bonds, certificates of deposit (CDs), and mutual funds.

A **stock** is a partial ownership of a company. The term **equities** is sometimes used for stock. The term means that the owner of a stock has an **equitable claim** on the company. Suppose a drug company called Curall issues 18,000,000 shares of stock. An investor who buys 18 **shares** of Curall stock owns $\frac{18}{18,000,000}$ or one millionth of the company.

EXAMPLE 1 Maria bought 100 shares of Curall at 12.6. She sold the shares three years later at 16.45. What is the difference between the price she paid for her shares of Curall and the price she received when she sold the shares?

Solution The number 12.6 means that one share of Curall cost $12.60 when Maria bought the stock. She paid:

$100 \times 12.6 = \$1,260$

She sold the stock for:

$100 \times 16.45 = \$1,645$

The difference between the selling price and the purchase price is:

$\$1,645 - \$1,260 = \mathbf{\$385}$

Stocks are usually **traded** (bought and sold) through a **stockbroker** (often called a **broker**) who charges a **commission,** a fee for buying or selling stocks. The fee is sometimes a flat charge plus a percent of the **gross.** The gross is the total amount of the **transaction** (either buying or selling) before the commission is calculated. Remember to add the commission to the cost of buying a stock and to subtract the commission from the selling price.

EXAMPLE 2 To sell her Curall stock, Maria used a broker who charged $17.00 + .006 × gross. Calculate the amount Maria will receive for her sale of 100 shares of stock.

Solution The gross amount of the sale was 100 × $16.45 = $1,645. The commission for the sale is:

.006 × $1,645 + $17.00 = $26.87

Maria will receive $1,645 − $26.87 = **$1,618.13** for selling her stocks.

Today it is possible to pay lower commission fees by buying and selling stocks on the Internet. The commissions for Internet transactions are almost always lower than brokers' fees.

The business pages of many newspapers show the daily activity of the stocks that are traded on major **exchanges.** The largest and most famous of these is the NYSE, the New York Stock Exchange. Other exchanges include the NASDAQ and the American Stock Exchange. There are also stock exchanges in many other cities around the world such as London, Tokyo, and Toronto.

Below is a typical line from a **stock listing** in a newspaper. The listing is for a company called Marlex that trades under the symbol MLX. (The company names in this lesson have been invented, but the numbers in the examples and exercises are from real companies.) Following the listing is an explanation of each number. Take the time to learn how these listings work. Then look in the business pages of a newspaper and read the stock listings of some companies that you are familiar with.

| 52-Week | | | | Yld | | | Sales | | | | |
High	Low	Stock	Div.	%	P/E	100s	High	Low	Last	Chg.
27.11	15.54	MLX	.30	1.2	26	5254	25.85	24.90	25.84	+0.26

- **27.11** is the **highest value** of the stock in the past year (52 weeks). In other words, one share of Marlex stock sold for $27.11.

- **15.54** is the **lowest price** of the stock in the past year.

- **MLX** is the **trading symbol,** a kind of abbreviation, of the stock's name, Marlex.

- **.30** is the **payment** in dollars that the company expects to pay for each share. This payment of $.30 is called a **dividend.** If there is a blank in the space under "Div.," the company does not pay a dividend.

- **1.2** is the **yield.** This is the annual dividend divided by the closing price (the next to the last number in the listing). The number is expressed as a percent. For Marlex, the yield is $0.30 divided by a price of $25.84 a share.

  $0.30 ÷ $25.84 = 0.0116 or 1.2%

- **26** is the **price-to-earnings** ratio. This is the price of one share divided by the dividend paid on each share. The number is not measured in dollars. The P/E ratio is based on the dividend paid over the past four quarters and cannot be calculated from the numbers in the listing.

- **5254** is the **total number of shares,** in hundreds, that were traded. For the day shown in the listing, 100 × 5,254 = 525,400 shares of Marlex stock were traded.

The last four numbers are measured in dollars:

- **25.85** is the **highest price of a share** of Marlex stock for the day.

- **24.90** is the **lowest price of a share** of Marlex for the day.

- **25.84** is the **last trading price,** called the **closing price,** for the day.

- **+0.26** is the **difference** between the last price of the stock on the previous day and the last price on the current trading day. Marlex was up $0.26 from the previous day's closing price.

A **bond** is a way for companies and government organizations to raise money. A bond is a loan from an investor to the company or government agency. The Treasury Department of the U.S. government sells **treasury bonds.** State and local governments sell **municipal bonds.** Companies sell **corporate bonds.**

A bond investor receives interest from the issuer of the bond at a **coupon rate.** The coupon rate is the percent the issuer of the bond is required to pay the holder.

EXAMPLE 3   A $500 bond pays $25 a year in interest. What is the coupon rate of the bond?

Solution   Divide the interest by the value of the bond.

$25 ÷ $500 = 0.05

Change 0.05 to a percent.

0.05 = **5%**

The coupon rate is **5%.**

A bond is a loan to a company or a government agency for a fixed amount of time. When a bond **matures,** in 1, 5, 10, or even 30 years, the investor gets back the original loan amount.

Stocks and bonds are sometimes lumped together under the term **securities.** The word *security* simply describes any stock or bond that can be traded.

A **certificate of deposit (CD)** is an investment with a bank. CDs are insured deposits that pay interest and require that the money remains invested for a fixed period of time.

A **mutual fund** is an investment company that uses the cash of investors to buy stocks and bonds in a particular way. The company's **prospectus** is a legal document that explains in great detail how an investor's money will be spent.

There are mutual funds that invest in specialized areas such as transportation or energy or natural resources. Other mutual funds buy only bonds. **Index funds** buy only the stocks that make up certain well-known standards such as the Dow Jones Industrial Average or the Standard & Poor's 500. When a standard like the S&P 500 is up for the year, the stocks held in an index fund usually gain by a similar percent.

The daily listings in business pages of newspapers include the **NAV** or **net asset value** of mutual funds. The NAV is similar to the share price of a stock. At the end of a business day, the fund company adds the value of all the investments in the fund and subtracts any fees or commissions. The resulting number divided by the total number of shares in the fund is the NAV.

Many Americans do most of their investing in their retirement accounts such as **IRAs**—Individual Retirement Accounts—**401(k)** plans, or **403(b)** plans. The complicated name 401(k) or 403(b) simply refers to a section of the Internal Revenue Code that describes the requirements for each saving plan. These retirement accounts allow investors to **defer** taxes on the amount they contribute to the accounts. This means that investors pay no income tax on the money they put in the accounts until they withdraw the money when they retire.

To solve the problems in the next exercise, review:

- decimals, pages 229–232
- finding a percent of a number, page 236
- finding what percent one number is of another, page 238

EXERCISE 8

Part A

Solve each problem. Use a calculator when needed.

1. Find the gross value of 40 shares of stock that trade at 38.4.

2. What is the coupon rate of a $2,000 bond that pays $80 interest in a year?

3. Marcos has 723.647 shares of a mutual fund in his retirement account. If the NAV of the fund is 15.90, what is the total value, to the nearest dollar, of Marcos' mutual fund?

Part B

Use the following table of commission rates to answer problems 4 to 7.

Internet Trades

Stocks	$19.95 per transaction
Mutual Funds	$19.95 per transaction

Broker-Assisted Stock Trades

under $500	$20 + .006 × gross amount
$500 – $1,000	$25 + .0025 × gross amount
$1,001 – $10,000	$35 + .002 × gross amount
over $10,000	$50 + .001 × gross amount

4. What is the total cost of a broker-assisted purchase of 80 shares of SureBet stock that is selling at 46.3?

5. Nurdan put $2,500 into a mutual fund on the Internet. What is the fee for her transaction?

6. Jeff used the Internet to sell 20 shares of Texoil stock that was trading at 32.45. What net amount did Jeff receive for the transaction?

7. Sandra used a broker to buy 60 shares of Goodeal stock that sold for 22.7. She used a broker again to sell the stock at 26.3. What was her net profit on the stock?

Part C

The table below shows the stock listings for five companies: Branscom (BRA), Cadmium Electronics (CAD), FastFinders (FSF), NorthAir (NAR), and Romax (RMX). Use the stock page below to answer problems 8 to 19.

| 52-Week | | | | Yld | | | Sales | | | | |
High	Low	Stock	Div.	%	P/E	100s	High	Low	Last	Chg.
10.15	4.00	BRA	...	...	...	120	8.27	7.40	7.75	−0.75
81.75	57.95	CAD	1.27	1.6	...	247	80.29	78.93	80.11	+0.98
15.26	7.20	FSF	.20	2.0	...	193	9.95	9.55	9.92	+0.44
30.15	11.59	NAR	.24	0.9	16	240	29.00	26.80	27.80	−1.30
37.69	26.00	RMX	2.50	7.2	19	1415	35.00	34.66	34.77	−0.03

8. Which of the five stocks has the highest closing price for the day?

9. Which of the five stocks has the lowest closing price for the day?

10. Which of the five stocks has the highest volume of sales for the day?

11. Which of the five stocks has the lowest volume of sales for the day?

12. What is the 52-week high for NorthAir?

13. Which of the five stocks pays the highest dividend?

14. How many shares of NorthAir were traded the day of the listings?

15. Which of the five stocks have a lower closing price than the previous day's closing price?

16. Which of the five stocks has a closing price that was nearly twice its lowest trading price of the year?

17. What is the difference between the highest price and the lowest price of Cadmium in the past year?

18. What is the previous day's closing price of a share of FastFinders?

19. The closing price of a share of NorthAir is how much less than its highest trading price for the year?

Post-Lesson Vocabulary Reinforcement

Read the definitions below. Then fill in the blanks with the letters of the appropriate terms from the list on the right.

_____ 1. the total of a transaction before the commission is calculated

_____ 2. an investment that represents a partial ownership of a company

_____ 3. a term sometimes used for *stocks*

_____ 4. the amount a company pays for each share of stock

_____ 5. a way for companies and governments to raise money

_____ 6. to not have to pay until later

_____ 7. net asset value

_____ 8. Individual Retirement Account

_____ 9. a loan from an investor to a company or government

_____ 10. a managed investment account that buys stocks and bonds

_____ 11. stocks or bonds that can be traded

_____ 12. a fee for buying or selling stocks, usually charged by a stockbroker

_____ 13. a legal document that explains how invested money will be spent

_____ 14. the percent a bond issuer is required to pay the holder

_____ 15. certificate of deposit

a. dividend

b. selling bonds

c. equities

d. defer

e. IRA

f. gross amount

g. stock

h. NAV

i. mutual fund

j. CD

k. prospectus

l. securities

m. coupon rate

n. bond

o. commission

Choose the appropriate words from the right to complete the sentences below.

16. Stocks are usually _____ (bought and sold) through a stockbroker.

17. _____ managers buy only the stocks that make up certain well-known standards.

18. The Treasury Department of the U.S. government sells _____.

19. _____ are deposits that require the money to remain invested for a period of time.

20. State and local governments sell _____.

21. Companies sell _____.

22. 401(k) and 403(k) are examples of _____.

23. The _____ is similar to the share price of a stock.

treasury bonds

traded

net asset value

municipal bonds

IRAs

corporate bonds

index fund

CDs

Language Builder

ACTIVITY A A **clause** is a kind of sentence inside another sentence. You've already practiced using "if clauses" in Lesson 6. Now you will practice recognizing and using other clauses that begin with the words *that, who, when, how, until,* and *whenever.* These words are called *clause markers* (CM). Notice that a subject (S) plus a verb (V) usually makes up a sentence.

Here is a sentence with a clause, taken from this lesson:

<u>*The term*</u> <u>*means*</u> **<u>*that*</u>** **<u>*the owner of a stock*</u>** **<u>*has*</u>** **<u>*an "equitable claim" on the company*</u>**.
    S   V   **CM**   **S**   **V**

Now examine the following sentences. <u>Underline</u> the clause or clauses in each. Then find the clause marker (CM), the complete subject (S), and the main verb (V) in each clause and write the letters beneath them as in the **boldfaced** examples above.

1. Maria received $385 when she sold her shares of Curall.

2. The number 12.6 means that one share of Curall cost $12.60 when Maria bought it.

3. Stocks are usually traded through a stockbroker who charges a commission.

4. The listing is for a company called Marlex that trades under the symbol MLX.

ACTIVITY B Find other examples of clauses in this lesson. Then write your own clauses to complete the following sentences. Use information from the lesson.

5. CDs are insured deposits that pay _____.

6. A stockbroker is a person who _____.

7. A commission is a fee that a stockbroker _____.

8. A prospectus explains how _____.

9. When I _____, I can defer the taxes until later.

10. I want to buy a stock that _____.

UNIT 2
BORROWING
MONEY

Lessons 9 to 13 examine the ways consumers pay more than they realize for the things that they buy. Many people end up borrowing money, even for a simple $5 item, every time they fail to pay the full amount of their credit card bills by the due date. These lessons introduce some of the hidden costs involved with borrowing money.

LESSON 9: INTEREST

Pre-Lesson Vocabulary Practice

Study the boldfaced terms in the following sentences. Then match the definitions on the right to the terms below.

_____ 1. **Substitute** $750 for the principal.

_____ 2. Change 9 months to a **fraction** of a year.

_____ 3. Jin **borrowed** $800 from his parents.

_____ 4. Jin **agreed** to pay them back.

_____ 5. Jin agreed **to pay them back.**

_____ 6. Compound interest is calculated at regular **intervals** such as every three months.

a. **specific periods of time**

b. **a part of**

c. **replace, put in its place**

d. **to return money**

e. **to take as a loan**

f. **consent, go along with**

With a partner, take turns choosing a term from the list and then giving the meaning.

Below is a review of vocabulary items from previous lessons. Read the definitions below. Then fill the blanks with the appropriate terms from the list.

_____ 7. to find a specific dimension or quantity

_____ 8. to use numbers to determine or figure something out

_____ 9. by the year, per year

_____ 10. to work for, to make money

_____ 11. an amount of money placed in a bank account

_____ 12. the amount of money remaining in a bank account

_____ 13. an organization or a company that deals with money, such as a bank

a financial institution

to calculate

annual

to measure

a deposit

to earn

the balance

Now try another review. Write the terms on the right in the appropriate blanks.

_____ 14. a formula

_____ 15. the first quarter of a calendar year

_____ 16. a fraction

_____ 17. percent

_____ 18. a decimal

January–March

0.04

%

$i = prt$

$\frac{1}{4}$

Interest

Interest is money that money makes. Interest is a payment for using another person's money. A bank pays a customer interest for using the customer's money. A customer pays interest for using the bank's money in a loan.

Interest is measured in dollars and calculated as a percent of **principal.** Principal is the amount of money on which interest is paid. Principal is the amount of savings in a customer's account in a bank or the amount of a loan.

The formula for calculating interest is:

interest = principal × rate × time

Principal is the money on which interest is paid.
Rate is the percent used to calculate the interest.
Time is a number of years or a fraction of a year.

Simple interest is money paid on only the principal.

EXAMPLE 1　Find the simple interest on $750 at 4% annual interest for 3 years.

Solution　Change 4% to a decimal: 4% = 0.04
Substitute $750 for the principal, 0.04 for the rate, and 3 for the time.

interest = $750 × 0.04 × 3 = **$90**

EXAMPLE 2　What is the new principal on $1,000 that has earned 6% annual interest for 9 months?

Solution　Change 6% to a decimal: 6% = 0.06
Change 9 months to a fraction of a year.

9 months = $\frac{9}{12} = \frac{3}{4}$ year

interest = $1,000 × 0.06 × $\frac{3}{4}$ = $45

Add the interest to the principal: $1,000 + $45 = **$1,045**

The formula for interest is $i = prt$. This formula expresses the relationship among interest, principal, rate, and time. If you know three of the amounts, it is easy to calculate the fourth amount.

To find the interest rate, solve for r in the formula: $r = \dfrac{i}{pt}$

To find the time, solve for t in the formula: $t = \dfrac{i}{pr}$

To find the principal, solve for p in the formula: $p = \dfrac{i}{rt}$

EXAMPLE 3

BANK DEPOSIT SLIP

CASH ▸	5 0 0 . 0 0
CHECKS ▸	.
▸	.
▸	.
SUBTOTAL ▸	5 0 0 . 0 0
LESS CASH RECEIVED ▸	– . – –
$	5 0 0 . 0 0

Martha deposited $500 in a savings account that earned simple interest. At the end of three years the account had a balance of $567.50. Find the rate of simple interest that she earned on the savings account.

Solution   Find the total amount of interest:

$567.50 – $500 = $67.50

Substitute $67.50 for i, $500 for p, and 3 for t in the formula $r = \dfrac{i}{pt}$.

$$r = \frac{\$67.50}{\$500 \times 3} = \frac{\$67.50}{\$1,500}$$

$$r = 0.045 = \mathbf{4.5\%}$$

EXAMPLE 4 Jin borrowed $800 from his parents and agreed to pay them back with 6% simple annual interest. Jin paid his parents a total of $872. How long did it take Jin to pay back the money he borrowed?

Solution   Find the total amount of interest: $872 − $800 = $72.

Substitute $72 for i, $800 for p, and 0.06 for r in the formula $t = \frac{i}{pr}$.

$$t = \frac{\$72}{\$800 \times 0.06}$$

$$t = \frac{\$72}{\$48} = \frac{3}{2} = 1\frac{1}{2} \text{ years or } \textbf{1 year and 6 months}$$

Compound interest is money paid on both the principal and the interest that has already been paid on the principal. Compound interest is calculated at regular intervals such as every three months (quarterly), every month, or even every day.

Not surprisingly, compound interest is complicated. Financial institutions use computer programs to make their calculations. However, to understand compound interest, follow the next example carefully.

EXAMPLE 5 What is the new principal at the end of one year on $2,000 at 8% annual interest if the interest is compounded quarterly?

Solution The interest for the first quarter is:
interest = $2,000 × 0.08 × $\frac{1}{4}$ = $40

The principal at the end of the first quarter is:
$2,000 + $40 = $2,040

The interest for the second quarter is:
interest = $2,040 × 0.08 × $\frac{1}{4}$ = $40.80

The principal at the end of the second quarter is:
$2,040 + $40.80 = $2,080.80

The interest for the third quarter is:
$2,080.80 × 0.08 × $\frac{1}{4}$ = $41.616 or $41.62

The principal at the end of the third quarter is:
$2,080.80 + $41.62 = $2,122.42

The interest for the final quarter is:
$2,122.42 × 0.08 × $\frac{1}{4}$ = $42.4484 or $42.45

The principal at the end of the final quarter is:
$2,122.42 + $42.45 = **$2,164.87**

Notice that for each quarter the new principal is multiplied by the same rate, 0.08 and the same time, $\frac{1}{4}$. You can simplify 0.08 × $\frac{1}{4}$ to 0.02.

Compound interest can be calculated using the steps above, however, using a computer program is much easier.

To solve the problems in the next exercise, review:

- multiplying decimals, page 231
- finding a percent of a number, page 236
- interest, page 238

EXERCISE 9

Part A

Solve each problem. Use a calculator and the formulas on page 95 of this lesson.

1. What is the simple yearly interest on $750 at $2\frac{1}{2}$ % annual interest?

2. Find the simple interest on $1,800 at 2.5% annual interest for eight months.

3. What is the new principal on $3,000 that earned $8\frac{1}{2}$ % annual interest for three years and six months?

4. Chan borrowed $500 from his brother. One year and three months after he borrowed the money, Chan paid his brother $562.50. What interest rate did Chan pay?

5. What is the simple interest on $2,000 at $5\frac{1}{4}$ % for four years?

6. To buy new furniture, Jane and Ike borrowed $1,600. They agreed to pay 14.5% annual interest. When they paid back their loan, they wrote a check for $1,716. For how long did they borrow the money?

7. Shirley borrowed money at $9\frac{3}{4}$ % annual interest. She paid back all the money including $195 in interest four months after she borrowed the money. How much did Shirley borrow?

8. After one and a half years, the account into which Bill deposited $900 had a total of $1,001.25. What interest rate did he earn on his deposit?

9. Find the simple interest on $3,600 at $4\frac{3}{4}$ % annual interest for nine months.

10. Imani borrowed $400 from her sister. Imani agreed to pay 7% simple annual interest. When she paid the money back, she gave her sister $470. For how long did Imani borrow the money?

> **Don't miss this great opportunity!**
>
> **2.5% interest on Savings Accounts!**
>
> Don't miss out – a rate this good won't last long!
>
> **Fixed rate for 6 months!**

Part B

11. Keiko put $5,000 in an account that earned 9% annual interest. If the interest was compounded once each year, how much was in Keiko's account at the end of three years?

12. Wu put $1,400 in an account that earned 3.5% annual interest. If the interest is compounded each month, how much will be in Wu's account at the end of three months?

Post-Lesson Vocabulary Reinforcement

Choose terms from the right to complete each sentence.

1. _____ is measured in dollars and calculated as a percent of principal.

2. _____ is the amount of money on which interest is paid.

3. _____ is money paid on only the principal.

4. _____ means every three months.

5. _____ is money paid on both the principal and the interest that has already been paid on the principal.

6. _____ pay their customers interest for using the customers' money.

Compound interest

Principal

Interest

Banks

Simple interest

Quarterly

Write the letters of the different terms from the formula on the appropriate blanks.

7. **a.** *interest =* **b.** *principal* $\times$ **c.** *rate* $\times$ **d.** *time*

$750 at 4% annual interest for 3 years = $90

___ ⬆   ___ ⬆   ___ ⬆ ___ ⬆

In the blanks, write *T* if the statement is true and *F* if the statement is false.

_____ **8.** Interest is money that money makes.

_____ **9.** A bank pays a customer interest for loaning a customer money.

_____ **10.** A customer pays interest for using the bank's money in a loan.

_____ **11.** Time is the percent used to calculate the interest.

_____ **12.** Rate is usually a number of years or a fraction of a year.

_____ **13.** The formula $i = prt$ expresses the relationship among interest, principal, rate, and time.

Now correct the false statements above on a separate sheet of paper.

Language Builder

ACTIVITY A This exercise will help you learn to correctly use **prepositions** in English. First, it's important to note <u>how</u> different prepositions are used. For example:

for: *Interest is a payment **for** using another person's money.* The preposition **for** in this sentence is used to express a reason or a purpose.

for: *Find the simple interest on $750 at 4% annual interest **for** 3 years.* The preposition **for** in this sentence is used to indicate an amount of time.

in: *Principal is the amount of savings **in** a customer's account **in** a bank or the amount of a loan.* The preposition **in** in this sentence is used to indicate location (a place).

of: *Principal is <u>the</u> amount **of** money on which interest is paid.* The preposition **of** in this sentence is used after <u>a</u> or <u>the</u> + a "quantity word" –for example, <u>the</u> amount.

on: *Simple interest is money paid **on** only the principal.* The preposition **on** in this sentence is used to talk about money.

Note: All of these prepositions may be used in other ways as well.

Write the appropriate preposition from the list above in each blank.

1. Interest is calculated as a percent _____ principal.

2. Find the simple interest _____ $750 at 4% annual interest for 3 years.

3. Martha deposited $500 _____ a savings account.

ACTIVITY B This is a review of the **passive:** the **verb *to be*** (*is / are*) + the **past participle** of the verb. Here's an example sentence from this lesson:

<u>Interest **is measured** in dollars and **calculated** as a percent of principal.</u>

In the blanks, check ✓ the sentences that contain the passive and <u>underline</u> the verb *to be (is / are)* + the past participle of the verb. If a sentence does not contain a passive form of the verb, do not check or underline anything. Number 4 is completed as an example.

4. ___✓___ Interest <u>is paid</u> on the principal.

5. _____ The principal of $1,000 has earned 6% annual interest for 9 months.

6. _____ Compound interest is calculated at regular intervals.

7. _____ Not surprisingly, compound interest is complicated.

8. _____ The percent used to calculate the interest is called "rate."

LESSON 10: LOANS

Pre-Lesson Vocabulary Practice

Study the terms listed on the right. Try to find them in the lesson and read the surrounding sentences to see if you better understand the meanings. Then work with a partner and take turns choosing a term from the list and giving the meaning.

Use the terms below in sentences of your own. Write them on a separate sheet of paper. Try to write original sentences to show that you know the meaning of the following terms. An example is provided for the first item.

1. routinely: *I routinely pay my bills on time.*

2. tuition

3. lender

4. borrower

5. installments

6. unpaid balance

7. short-term loans

8. dependable

9. insurance policies

10. due

Compare your work with a partner's and then with the class.

routinely – regularly, usually

tuition – the amount of money it costs to attend college

lender – the person or institution that lends or loans money, such as a bank

borrower – the person or institution that borrows money

installments – separate, partial payments

unpaid balance – the amount that the borrower still owes

short-term loans – loans made for short periods of time, such as 30 days

dependable – reliable

insurance policies – protection for property in case it is damaged or stolen

due – required to be paid

Loans

People routinely borrow money to make major purchases such as a car or a house, to pay college tuition, or to have extra cash for a vacation. Money that is borrowed is called a **loan.** The length of time the borrower has to pay back a loan is called the **term** of the loan. Almost always, the **lender** earns interest on the principal. To solve the problems in this lesson you will need the formulas in the previous lesson.

Borrowers pay back many loans in **installments.** Each installment, or payment, usually includes both principal and interest. A loan in which each installment is the same is called a **level-payment loan.** One of the most common types of loans is the mortgage loan for the purchase of a house. You will learn more about mortgage loans later in this unit.

EXAMPLE 1   Sandra borrowed $5,000 at 6% annual interest to buy a used car. She agreed to pay back the money in equal installments over four years. Calculate the amount of each monthly payment.

Solution   The total interest Sandra will pay is:

$i = prt = \$5,000 \times 0.06 \times 4 = \$1,200.$

The total that Sandra will have to pay is:

$\$5,000 + \$1,200 = \$6,200$

Divide the total by the number of months in four years ($4 \times 12 = 48$).

$\$6,200 \div 48 = \$129.166 \ldots$ or **$129.17**

EXAMPLE 2 Silvia borrowed $1,500 from a friend and paid back $1,882.50 three years later. What rate of interest did she pay on her loan?

Solution Silvia paid $1,882.50 − $1,500 = $382.50 in interest.

To find the interest rate, use the formula $r = \frac{i}{pt}$ from the last lesson.

$$r = \frac{\$382.50}{\$1,500 \times 3} = \frac{\$382.50}{\$4,500} = 0.085 = \textbf{8.5\% interest}$$

A **promissory note** is an agreement between an institution and an individual or between two people. The lender agrees to lend a specific amount of money to the borrower for a specific amount of time. An agreement in which the lender charges interest on the money is called an **interest-bearing note.**

EXAMPLE 3 Juan borrowed $1,000 at 4% annual interest from his brother Marco for eighteen months. What total amount, including interest, does Juan owe his brother?

Solution Eighteen months = 1 year and 6 months

$$= 1\tfrac{1}{2} \text{ or } 1.5 \text{ years}$$

The interest is:

$$i = prt = \$1,000 \times 0.04 \times 1.5 = \$60$$

Juan owes a total of $1,000 + $60 = **$1,060.**

On some loans the borrower pays interest on only the unpaid balance. With these loans, each monthly payment is different.

EXAMPLE 4 Marcia borrowed $2,000 at 9% annual interest for six months.
Calculate her first two monthly payments.

Solution   The interest due the first month is:

$i = prt = \$2,000 \times 0.09 \times \frac{1}{12} = \$15$

The amount of principal that is due each month is:

$\$2,000 \div 6 = \$333.33$

The total principal and interest for the first month is:

$\$333.33 + \$15 = \mathbf{\$348.33}$

The balance of the principal for the second month is:

$\$2,000 - \$333.33 = \$1,666.67$

The interest due the second month is:

$i = prt = \$1,666.67 \times 0.09 \times \frac{1}{12} = \$12.50$

The total principal and interest for the second month is:

$\$333.33 + \$12.50 = \mathbf{\$345.83}$

Banks make **short-term loans** for terms of 30 days, 60 days, or
90 days. These loans are often **non interest-bearing loans.** The bank
deducts the amount of interest in advance. The borrower receives the
amount he has requested minus the interest. This new amount is called
the **proceeds** of the loan. To simplify calculations, banks often think of
a year as twelve months, each with 30 days, or a total of 360 days.

EXAMPLE 5 Ramona took out a short-term loan of $1,250 from her bank
at 13% annual interest for 90 days. Find the proceeds that
Ramona will receive from the bank.

Solution   The time is 90 days $= \frac{90}{360} = \frac{1}{4}$ year.

The interest is:

$i = prt = \$1,250 \times 0.13 \times \frac{1}{4} = \$40.625$ or $\$40.63$

The proceeds are the total loan minus the interest:

$\$1,250 - \$40.63 = \mathbf{\$1,209.37}$

The legal documents that describe the details of loans contain specialized vocabulary that may seem intimidating. For example, interest rates are given **per annum.** This Latin phrase means *per year.* Some lending institutions list **APRs.** These letters stand for "**annual percentage rate.**"

Interest rates are sometimes described as one or two percentage points above the **prime.** The *prime rate* is the interest rate that banks charge their best, most dependable customers. If the prime is 4%, an interest rate of prime + 2% is simply 6%.

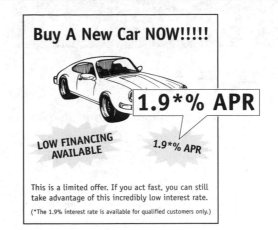

Buy A New Car NOW!!!!!

1.9*% APR

LOW FINANCING AVAILABLE   *1.9*% APR*

This is a limited offer. If you act fast, you can still take advantage of this incredibly low interest rate.

(*The 1.9% interest rate is available for qualified customers only.)

A **collateral loan** means that the borrower offers something of value that the lender can keep if the borrower fails to repay the loan. On a car loan, the car itself is the collateral. Stocks, bonds, insurance policies, a house, or a piece of land can serve as collateral. Collateral loans are sometimes called **demand loans** or **demand notes.** Legally, the lender can demand to get money back from the borrower. If the borrower fails to pay, the lender gets the collateral. The mathematics of collateral loans is no different than that of other loans.

EXAMPLE 6   Benji borrowed $1,500 on a demand note from his bank at a rate of 14% per annum. He paid the bank $1,815 on the day the note was due. Find the time period for the loan.

Solution   Benji paid interest of $1,815 − $1,500 = $315.

$$t = \frac{i}{pr} = \frac{\$315}{\$1,500 \times 0.14} = \frac{315}{210} = \textbf{1.5 years or 18 months}$$

To solve the problems in the next exercise, review:

- finding a percent of a number, page 236
- interest, page 238

EXERCISE 10

Part A

Solve each problem. You may use a calculator and the following formulas:

$$i = prt \qquad r = \frac{i}{pt} \qquad t = \frac{i}{pr} \qquad p = \frac{i}{rt}$$

1. Richard borrowed $500 from his sister at 10% annual interest for nine months. How much interest did he owe?

2. What total amount did Richard, in problem 1, pay back to his sister?

3. Suzanne borrowed $3,000 from a friend. Twenty-four months later she paid back $3,690. What interest rate did she pay on her loan?

Part B

Use the following information to answer problems 4 to 6.

SITUATION

To buy a used car, Carlos borrowed $6,500 at 8.5% per annum interest for four years.

4. What total amount of interest will Carlos have to pay on his loan?

5. Find the total amount, including interest, that Carlos will pay.

6. If he pays off his loan in equal monthly installments, how much will Carlos pay each month?

Part C

7. Ernesto's bank approved a 60-day short-term loan of $800 at 9% annual interest. What are the proceeds that Ernesto will receive on the loan?

8. Cleo got proceeds of $1,975 on a short-term loan of $2,000 at 15% per annum. How much time does Cleo have to pay back the loan?

9. Alejandro received proceeds of $878.25 on a 60-day short-term loan of $900. What annual interest rate did the bank charge him?

10. Connie borrowed $3,000 at 18.5% annual interest for two years. If she pays the loan back in equal installments, how much will she have to pay each month?

Part D

Use the following table to answer problems 11 to 17.

Car Loan Rates			
New Vehicles		Used Vehicles	
		less than 6 years old	6 years old or older
term	rate	rate	rate
36 months	7.25%	8.25%	9.5%
48 months	7.5%	8.5%	N/A
60 months	7.75%	8.75%	N/A

11. What is the interest rate for a four-year loan on a new vehicle?

12. What is the interest rate on a three-year loan for a used car that is 8 years old?

13. What is the interest rate on a five-year loan for a used car that is two years old?

14. Is it possible to get a four-year loan for a 10-year-old vehicle from this company?

15. Helen borrowed $16,000 for five years to pay for her new car. In total, how much interest will she pay on her loan?

16. Iwai borrowed $4,500 for three years to buy a 3-year-old car. How much will he pay for the car in all?

17. Anna and Tom borrowed $18,000 for three years to buy a new car. Calculate the amount of their monthly payment.

Part E

Use the following information to answer problems 18 to 20.

SITUATION

Mrs. Rivera borrowed $2,400 at 13.5% annual interest for six months. She agreed to make monthly installments that will include principal and interest on only the unpaid balance.

18. How much will Mrs. Rivera pay each month toward the principal?

19. Find the amount, including interest, of her first month's payment.

20. Find the amount, including interest, of the second month's payment.

Post-Lesson Vocabulary Reinforcement

Match the definitions with the terms. Write the appropriate letters in the blanks.

_____ 1. agreement between an institution and an individual or between two people

_____ 2. money that is borrowed

_____ 3. total amount of a loan minus the interest

_____ 4. Annual Percentage Rate

_____ 5. sometimes called "demand loans" or "demand notes"

_____ 6. the interest rate that banks charge their best, most dependable customers

_____ 7. agreement in which the lender charges interest on the money

_____ 8. loan in which each installment is the same

a. a loan

b. a level-payment loan

c. a promissory note

d. an interest-bearing note

e. APR

f. the prime rate

g. collateral loans

h. the proceeds

Choose the appropriate terms from the box to complete the sentences.

car loan	**major purchases**	**interest rates**
collateral	**mortgage loan**	**term of the loan**

9. People routinely borrow money to make _____ such as to buy a car or a house or to pay college tuition or to have extra cash for a vacation.

10. One of the most common types of loans is the _____ for the purchase of a house.

11. _____ are sometimes described as one or two percentage points above the prime.

12. On a _____, the car itself is the collateral.

13. Stocks, bonds, insurance policies, a house, or a piece of land can serve as _____.

14. The length of time the borrower has to pay back a loan is called the _____.

Language Builder

ACTIVITY A The **future tense with *will*** is used frequently in this lesson and throughout this book. You should understand how it works and when to use it. This sentence from the lesson shows that, for the future tense, the verb that follows *will* is always in the simple or infinitive form (*learn*). It also includes a future word *later*.

> You **will** learn more about mortgage loans <u>later</u> in this unit.

Often a future word is not included because a time in the future is not specified:

> *The total interest Sandra **will** pay is $1,200.*

In the following sentences, <u>underline</u> the parts of the future tense with *will*—the subject, verb, and *will*. The first item has already been done.

1. The total that <u>Sandra</u> <u>will</u> <u>have</u> to pay is $6,200.

2. Find the proceeds that Ramona will receive from the bank.

3. What total amount of interest will Carlos pay on his loan?

4. What are the proceeds that Ernesto will receive on the loan?

5. In total, how much interest will Helen pay on her loan?

ACTIVITY B Now put the following words in the correct order to form sentences with the future. Write your answers on a separate sheet of paper.

6. pay. including interest, will Find the total amount, have to that Carlos

7. including interest, What total amount, Juan the lender? will owe

8. will on their loan? Anna and Tom How much interest pay

9. formulas to solve You in this lesson. problems will need

ACTIVITY C In Lesson 7, you learned about **W/H information questions** that begin with a W/H question word such as <u>What</u> . . . or <u>How</u> . . . Below you will review them and practice their use with different verb tenses by putting the words in correct sentence order. Write the sentences on a separate sheet of paper.

10. the interest rate is on a new vehicle? for a four-year loan What

11. for the car in all? will pay How much Iwai

12. Silvia on her loan? did pay rate of interest What

13. his sister? Richard How much owe interest did

14. is What for a used car? on a three-year loan the interest rate

LESSON 11: INSTALLMENT BUYING

Pre-Lesson Vocabulary Practice

On the right is a list of terms from this lesson. Read the terms and their definitions. Then work with a partner to use each term in an original sentence.

Use the pictures below to help you understand the meanings of the various store items. Then find and read the words and their sentences in this lesson.

personal computer

digital camcorder

12-speed mountain bike

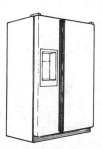

side-by-side refrigerator

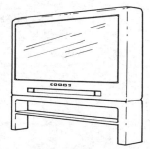

high-definition television

used pickup truck

installment buying – paying for something over a period of time

manufactured products – items that are produced or made in a factory

over a fairly long – during a long period of time

household appliances – items used in the home that are operated by electricity

dealers – persons who sell items

expensive items – store merchandise that costs a lot of money

rather than – instead of or versus

the full purchase price –the total amount or cost

all at once – completely

buyer – the person who buys or purchases something

as soon as – at the time

advanced to – loaned to

listed at/for – stated, as in an advertisement

Installment Buying

Economists use the term **durables** or **durable goods** to describe manufactured products that can be used over a fairly long period of time. Cars, trucks, and household appliances are examples of durable goods.

Stores and dealers that sell durable goods usually offer their customers the option of paying for expensive items by using an **installment plan** rather than paying the full purchase price all at once. An installment plan is an agreement between a dealer and a consumer. The consumer usually makes a **down payment** and then regular monthly payments. Each monthly payment includes part of the **balance due** as well as **interest** that the dealer collects for allowing the buyer to pay over a period of time. Although the buyer can use the item as soon as he or she agrees to the terms of the installment plan, the dealer actually owns the item until the last payment is made.

EXAMPLE 1   Alicia wants to buy a new computer listed for $709.97. She decides to buy the computer on an installment plan that requires a down payment of $75 and 12 monthly payments of $59 each. What total price will she pay for the computer?

Solution   Find the total cost of the down payment plus the 12 monthly payments.

$75 + (12 × $59) = $75 + $708 = **$783**

It is a good idea to calculate the interest rate on the monthly payments of an installment purchase. The dealer in the first example is lending Alicia 12 × $59 = $708 toward the purchase of her computer.

EXAMPLE 2 What is the interest rate on the installment plan described in Example 1?

Solution The interest is the difference between the total cost and the list price:

$783 − $709.97 = $73.03

Use the formula $r = \dfrac{i}{pt}$ to find the rate of interest.

The principal is the $708 that is advanced to Alicia, and the time is 1 year.

$$r = \frac{i}{pt} = \frac{\$73.03}{\$708 \times 1} = 0.1031\ldots \text{ or } \mathbf{10.3\%}$$

To solve the problems in the next exercise, review:

• finding what percent one number is of another, page 238

• interest, page 238

EXERCISE 11

Part A

Use a calculator to solve the following problems. Calculate each percent rate to the nearest tenth of one percent.

1. Max bought a color TV listed at $399. He is paying on an installment plan that requires a down payment of $80 and twelve monthly payments of $30 each. What total price will Max pay for the television?

2. Find the interest rate on the monthly payments in problem 1.

3. A digital camcorder has a list price of $499.99. Maki purchased the camcorder on an installment plan with a down payment of $100. She agreed to pay an additional $468 in six equal monthly installments. How much did she have to pay each month?

4. What was the interest rate on the installment payments in problem 3?

5. Pedro bought a 12-speed mountain bike that cost $598. He paid $75 down and made nine monthly payments of $65 each. Find the total price he paid for the bike.

6. What interest rate did Pedro, in the last problem, pay on the installments?

7. The Madisons bought a washer and dryer listed at $679. They made a down payment of $100 and 18 monthly payments of $36 each. What total price did they pay for the washer and dryer?

8. Find the interest rate on the Madisons' installment plan in problem 7.

Part B

9. Mr. and Mrs. Chung bought a 26-cubic-foot side-by-side refrigerator listed for $749.99. They made a down payment of $100 and paid an additional $711 in nine equal monthly payments. Find the amount of each monthly payment.

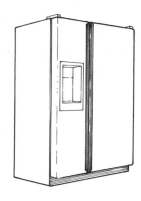

10. What interest rate did the Chungs pay on their installment plan?

11. Meaghan bought a suite of living room furniture listed at $820. According to her installment plan, she had to make a down payment of $125 and twelve monthly payments of $66 each. Find the total price she paid for her furniture.

12. Find the interest rate for the purchase of the furniture in problem 11.

13. Lisa wants to buy a 54-inch high-definition television listed at $1,399. If she buys on an installment plan, she will have to put down $200 and make 24 monthly payments of $59 each. What total price will she pay for the television?

14. Find the interest rate on Lisa's installment plan in problem 13.

Part C

15. Mohamed plans to buy a used pickup truck that is listed for $16,750 on an installment plan. He will put down $3,000 and make monthly payments of $255 for six years. What total price will he pay for the truck?

16. What is the interest rate on the monthly payments for Mohamed's truck?

17. What percent of the list price is the total that Mohamed paid for the truck in problem 15?

Post-Lesson Vocabulary Reinforcement

Write the terms from the right in the appropriate blanks in the paragraphs.

Economists use the term *durables* or **1.** _____ to describe manufactured **2.** _____ that can be used over a fairly long period of time. Cars, trucks, and household appliances are **3.** _____ of durable goods.

Stores and dealers who sell durable goods usually offer their customers the option of **4.** _____ expensive items by using an **5.** _____ rather than paying the full purchase price all at once. An installment plan is an **6.** _____ between a dealer and a consumer. The consumer usually makes a **7.** _____ and regular monthly payments. Each monthly payment includes part of the **8.** _____ as well as interest that the dealer **9.** _____ for allowing the buyer to pay over a period of time. Although the buyer can use the item as soon as he agrees to the terms of the installment plan, the dealer actually **10.** _____ the item until the last payment is made.

paying for

examples

owns

agreement

down payment

collects

balance due

durable goods

installment plan

products

In the blanks under the appropriate pictures, write the correct term from the list.

11. _____ **12.** _____ **13.** _____

personal computer

digital camcorder

12-speed mountain bike

side-by-side refrigerator

high-definition television

used pickup truck

14. _____ **15.** _____ **16.** _____

Language Builder

ACTIVITY A Here are more examples of **prepositions.** The first one is a review from Lesson 10.

on: *Find the interest rate **on** the monthly payments in the last problem.*
   The preposition **on** is often used to talk about money.

over: *Installments allow the buyer to pay for a purchase **over** a period of time.*
   The preposition **over** is often used before an expression of time.

between: *An installment plan is an agreement **between** a dealer and a consumer.*
   The preposition **between** is often used before two <u>nouns</u> connected with the word *and*—"... a <u>dealer</u> *and* a <u>consumer</u>."

Write the appropriate preposition from above in each blank.

1. The interest is the difference _____ the total cost and the list price.

2. Installment buying involves making an agreement _____ a buyer and a seller.

3. What interest rate did Pedro pay _____ the installments?

4. Each monthly payment includes interest that the dealer collects for allowing the buyer to pay _____ a period of time.

ACTIVITY B The **present continuous** tense of the verb consists of a form of the verb **to be** (*is / are*) + the present participle (the *-ing* form) of the main verb. This tense expresses an action that has <u>not</u> been completed. For example: *Max <u>is paying</u> for his new color TV on an installment plan* means Max has not finished paying for his new color TV. The *paying* has not been completed.

Review the simple present and past tenses and practice the present continuous tense by identifying the time of the action for the following sentences. Write the correct letter in each blank.

   A. The action is going on right now.

   B. The action is in the simple present tense—a simple statement of fact.

   C. The action is in the past—it has been completed.

5. _____ Alicia wants to buy a new personal computer.

6. _____ Max bought a color TV listed at $399.

7. _____ Mohamed plans to buy a used pickup truck.

8. _____ Pedro is paying more than $600 for his new mountain bike.

ACTIVITY C Now use each of the three tenses in sentences of your own. You can use the same verbs as in the exercise above. Write your sentences and compare them with a partner and with the entire class.

LESSON 12: CREDIT

Pre-Lesson Vocabulary Practice

Below are definitions for some of the boldfaced terms in this lesson. Read the definitions. Then, after finding them in the lesson, match the definitions with the terms in the margin.

_____ 1. plastic card used as a temporary substitute for paying for something

_____ 2. credit cards that stores issue for use in their specific locations

_____ 3. plastic card used for automatically deducting money from a customer's bank account

_____ 4. provide, give out

a. credit card

b. bank/debit card

c. store charge card

d. issue

Below is a list of terms from this lesson. Read the terms with a partner. Then, after finding them in the lesson, work with a partner to define each term and use it in an original sentence.

Term	Definition
5. retailers/retail stores	_____
6. widely	_____
7. money order	_____
8. finance charges	_____
9. credit card companies	_____
10. cash advances	_____
11. to fall behind	_____
12. restrict	_____
13. theft	_____
14. identity	_____
15. reputable	_____

Credit

Credit is a temporary substitute for money. Consumers often use **credit cards, bank cards** (sometimes called **debit cards**), and **store charge cards** instead of cash to make purchases. Retail stores and store chains **issue** plastic cards that can be used at a main store, at branches of the store, or occasionally at other retailers that have financial agreements with the main store. Major credit cards such as Visa, MasterCard, Discover, and American Express are widely accepted. To pay for purchases that have been made with a credit card, a consumer must write a check or send a money order to the bank or other financial institution that issued the card.

There are small differences among these cards. Some credit cards require that users pay the entire **balance** that appears on each month's statement. More commonly, customers are required to make a **minimum payment** each month. Some credit cards charge an **annual fee;** others do not. Some retail stores offer their customers **regular charge accounts** that allow customers to pay for their purchases within one month. There is usually no fee for these accounts. Retail stores also offer **revolving charge accounts** that require a minimum payment and charge interest on any **unpaid balance.**

The **finance charges** on unpaid balances vary widely. The annual interest rate of these charges can be as low as 6% or as high as 20%. Most credit card companies charge higher rates for cash advances than for regular purchases.

The **monthly statement** from a credit card company lists the **activity** in the account since the last statement. The activity includes payments, new purchases, and any cash advances. The table below lists the monthly activity in Anne's credit card account.

ACCOUNT SUMMARY

Activity since last statement

Date	Description	Amount $
6/20	payment received	−150.00
6/18	Century Emporium	83.70
6/25	Javier's Car Repair	102.49
6/26	Valley Online Service	16.92

EXAMPLE 1 For the statement listed above, what was the total of Anne's new purchases?

Solution   Add the last three amounts on the statement.

$$\$83.70 + \$102.49 + \$16.92 = \textbf{\$203.11}$$

A monthly credit card statement often includes a kind of mathematical formula for calculating charges and **amounts due.** Look at this summary in Anne's statement.

Account Summary				
	(−)	(+)	(+)	(=)
Previous Balance	Payments & Credits	Purchases & Advances	Finance Charge	New Balance
$409.27	$150.00	$203.11	$ _____	$ _____

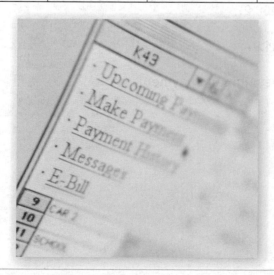

EXAMPLE 2 Use the summary in Anne's credit card statement to calculate the unpaid balance for the month.

Solution **The unpaid balance is the previous balance minus the payment.**

$409.27 − $150.00 = **$259.27**

EXAMPLE 3 The finance charge rate for Anne's credit card account is 18% per annum. What is the monthly finance charge on the unpaid balance in Anne's account?

Solution **The monthly finance charge rate is 18% ÷ 12 = 1.5%. Find 1.5% of $259.27.**

0.015 × $259.27 = $3.88905 or **$3.89**

EXAMPLE 4 Find the new balance for Anne's monthly credit card statement.

Solution

previous balance	$409.27
payment	−150.00
	$259.27
new purchases	+203.11
	$462.38
finance charge	+ 3.89
new balance	$466.27

EXAMPLE 5 Anne's credit card company requires a minimum payment that is either 7.5% of the new balance or $20, whichever is larger. What minimum payment does Anne have to pay for the month?

Solution **Find 7.5% of the new balance, $466.27.**

0.075 × $466.27 = $34.97025 or **$34.97**

Since $34.97 is more than $20, the minimum payment is $34.97.

Credit cards are popular in the U.S. For some people they are too popular. It is easy to pull a plastic card from a wallet and say, "Charge it." It is also easy to fall behind with payments or to make only minimum payments. Credit card companies have the right to impose **late charges** and to raise the interest rate they charge on unpaid balances as soon as a customer is late with a payment. Although some states have laws that limit the interest rate that can be charged on loans, some states do not have such laws. Large credit card companies are in states that do not have laws that restrict interest. The laws in the state where a customer makes a purchase have nothing to do with the interest rate of his or her finance charge.

Credit card theft, often called **identity theft,** is also a problem. Many people lose money because they give their credit card numbers to strangers over the phone. Be careful when you give your credit card number over the phone. Always be sure that you give your credit card numbers only to reputable businesses.

Note: The last problem in the next exercise shows the risk in paying only minimum payments on a credit card account.

To solve the problems in the next exercise, review:

- finding a percent of a number, page 236
- interest, page 238

Part A

Use a calculator to solve these problems.

SITUATION

Franco has a credit card with a company that charges 11.9% annual interest on unpaid balances and requires a minimum payment of $20 or 15% of the unpaid balance, whichever is greater. Use the following summary statement to answer problems 1 to 4.

ACCOUNT SUMMARY

Previous Balance		$795.23
Payments (2/15)		−$250.00

Purchases

2/12	Discount Drugs	$32.06
2/19	Uptown Butcher	$28.79
2/19	Smith's Sports	$56.28
2/24	Tommy's Tunes	$43.01

1. What is the total of the new purchases for the month?

2. Find the finance charge on the unpaid balance.

3. Calculate the new balance in Franco's account.

4. What is the required minimum payment for the month?

Part B

SITUATION

The revolving charge card account that Renata uses at the Century Emporium charges 16.5% annual interest on unpaid balances and requires a minimum payment of $20 or 6.5% of the unpaid balance, whichever is greater. Use the following summary of Renata's monthly Century Emporium account to answer problems 5 to 8.

ACCOUNT SUMMARY

Previous Balance		$1,073.88
Payments (5/16)		−$300.00

Purchases

5/10	sportswear	$89.97
5/13	bedding	$320.19
5/13	toys	$18.49
5/20	garden shop	$72.98

5. Find the total of the new purchases in Renata's account.

6. What is the finance charge on last month's unpaid balance?

7. Find the new balance in Renata's revolving charge account.

8. What is the minimum payment that Renata will have to make for her May statement?

Part C

SITUATION

Mrs. Olmstead's credit card company charges a rate of 18% annual interest on unpaid balances and requires a minimum payment of $20 or 2.5% of the unpaid balance, whichever is greater. Use the following summary to answer problems 9 to 12.

ACCOUNT SUMMARY

Previous Balance		$246.81
Payments (10/12)		−$40.00

Purchases

10/4	Jet Green	$216.00
10/9	Cat Boutique	$14.25
10/15	Gail's Books	$39.56
10/19	Civic Auditorium	$45.00
10/19	Acme Car Rental	$89.91

9. What is the total of this month's purchases?

10. What is the finance charge on the unpaid balance?

11. Calculate the new balance for the month.

12. What minimum payment does Mrs. Olmstead have to make on her account?

Part D

SITUATION

Joel put a charge of $2,000 on his credit card. The credit card company charges an annual rate of 18%, or 18% ÷ 12 = 1.5% per month on any unpaid balance. His credit card company allows a small minimum payment of 2.5% of the balance due.

13. Find the minimum payment that Joel made the first month.

14. If Joel pays only $50 each month and makes no additional purchases, his account for three months will have the following balances.

Month 1 $2,000.00

             −   50.00

Month 2 $1,950.00   × 0.015 = $29.25 finance charge

             +   29.25

             $1,979.25 balance due

             −   50.00

Month 3 $1,929.25   × 0.015 = $28.94 finance charge

             +   28.94

             $1,958.19 balance due

             −   50.00

             $1,908.19 unpaid balance after third payment

Calculate the monthly finance charges and the balances that will be due if Joel continues paying only $50 a month for one full year. Then compare the amount he will have paid in one year to the amount he will still owe.

Post-Lesson Vocabulary Reinforcement

Choose terms from the margin and write them in the appropriate blanks.

1. _____ is a temporary substitute for money.

2. _____ such as Visa, MasterCard, and American Express are widely accepted.

3. _____ allow customers to pay for their purchases within one month, usually without a fee.

4. Credit card theft, often called _____, is a problem.

5. _____ require a minimum payment and charge interest on any unpaid balance.

revolving charge accounts

credit

identity theft

regular charge accounts

credit cards

Write the letters of the terms on the appropriate blanks.

_____ 6. $409.27

_____ 7. $\frac{-150.00}{\$259.27}$

_____ 8. $\frac{+203.11}{\$462.38}$

_____ 9. $\underline{+3.89}$ (1.5%)

_____ 10. $466.27

a. **finance charge**

b. **new purchases**

c. **new balance**

d. **previous balance**

e. **payment**

In the blanks, write *T* if the statement is true and *F* if the statement is false. If a statement is false, work with a classmate to rewrite the statement, making it true.

_____ 11. Credit cards are sometimes called debit cards.

_____ 12. Customers with credit cards are commonly required to make a minimum payment each month.

_____ 13. The monthly statement from a credit card company lists the activity in the account before the last statement.

_____ 14. For some people, credit cards are too popular.

_____ 15. The finance charges on unpaid balances do not vary widely.

Language Builder

ACTIVITY A **Modals** are words that are used with verbs to express a particular meaning. The modal always has the same form, and the verb is always in the simple or infinitive form. Two modals used in this lesson are *can* and *will*. One use of the modal *can* is to express <u>possibility</u>, as in *plastic cards can be used at a store*. The modal *will* was introduced in Lesson 10. In this lesson, it is used to indicate an action that will happen in the <u>future</u>, as in *his account will have the following balances*.

Put the words, based on sentences from this lesson, in the correct sentence order. Use a separate sheet of paper.

1. as low as   can   6%.   The annual interest rate   be

2. will   on Renata's   the finance charge   be   unpaid balance   What   ?

3. next month.   be   That balance   due   will

4. loans.   be charged   can   A high interest rate   on some

5. still   At the end of the year,   Joel   owe $1,908.19.   will

6. instead of   be used   Credit cards   cash   can   in many stores.

Now make up sentences of your own with *can* and *will* and share them with a classmate and with the entire class.

ACTIVITY B **Comparative adjectives** were introduced in Lesson 4. Remember that the regular form of these adjectives ends in *-er* and that they are used to compare two things. Many times they are followed by the word *than,* as in *Most credit card companies charge <u>higher</u> rates for cash advances <u>than</u> for regular purchases.*

If there are comparative adjectives in the following sentences, underline them.

7. Anne's credit card company requires a minimum payment that is either 7.5% of the new balance or $20, whichever is larger.

8. Joel checked to see if another retailer would accept his credit card.

9. Mrs. Olmstead's credit card company charges a rate of 18% annual interest on unpaid balances and requires a minimum payment of $20 or 2.5% of the unpaid balance, whichever is greater.

10. To pay for purchases made with a credit card, a consumer must write a check or send a money order to the bank or other financial institution that issued the card.

ACTIVITY C Now work with a partner and make up sentences of your own that contain the following comparative adjectives. Then compare your work with other partners. Below are samples of comparative adjectives.

greater   *higher*   *larger*

LESSON 13: MORTGAGES

Pre-Lesson Vocabulary Practice

Study the new terms on the right. Then try to find them in the lesson and read the surrounding sentences to see if you better understand the meanings.

With a partner, take turns choosing a term from the list and then giving the meaning.

Now try to use the terms in sentences of your own. Try to write original sentences to show that you know the meaning of the following terms. An example is provided for the first item.

1. condominium: *I want to buy a condominium, not a house.*

2. sizeable: _____

3. prospective buyer: _____

4. put down: _____

5. asking price: _____

6. pay off/paid off: _____

7. long-term mortgage: _____

8. short-term mortgage: _____

9. goes toward: _____

Compare your work with a partner's and then with the rest of the class.

Write the words on the right in the appropriate blanks.

Phil and Patty Sanders' 20-year mortgage is for

$65,000	**10.** _____	the rate
× 8.5%	**11.** _____	the principal
× $\frac{1}{12}$ of a year	**12.** _____	the first month's interest payment
= $460.42	**13.** _____	the time

a condominium (condo) – a dwelling, like an apartment, located in a building where there are usually other condominiums

sizeable – rather large

prospective buyer – person planning to buy

put down – initially pay

asking price – amount of money the seller expects to be paid

pay off/paid off – pay/paid for completely

long-term mortgage – a mortgage that does not have to be paid off for many years

short-term mortgage – a mortgage that has to be paid off within a given number of years

goes toward – is paid on/for

Mortgages

An installment loan for the purchase of a house, condominium, or apartment is called a **mortgage.** The time period for a mortgage is usually many years. A 30-year mortgage is common. The **collateral** for a mortgage is the property itself. In fact, the lender—often a bank— actually owns the house until the buyer has paid off the mortgage. Every month the borrower makes a payment that includes both interest and principal to the lender.

When purchasing a house, buyers usually make a sizeable **down payment.** Mortgage lenders like prospective buyers to put down at least 20% of the asking price of a property.

The lender charges either a fixed rate of interest or an adjustable rate of interest. For a **fixed-rate mortgage,** the interest rate never changes for the term of the mortgage. For an **adjustable-rate mortgage** (ARM), the interest rate changes every year. The amount that an adjustable-rate mortgage can change each year is usually limited to one or two percentage points. Also, the maximum to which the interest rate can rise is usually limited. This upper limit is called the **cap.**

The time period of a mortgage affects the total amount that a borrower will pay. A long-term mortgage has the advantage of lower monthly payments, but the borrower pays more money over the term of the mortgage. A short-term mortgage has higher monthly payments, but the borrower pays off the debt more quickly.

SITUATION

Phil and Patty Sanders have a 20-year mortgage for $65,000 at 8.5% annual interest. Their monthly payment is $564.09.

EXAMPLE 1 How much of the first month's payment is for interest?

Solution Use the formula $i = prt$. The principal is $65,000, the amount of the mortgage.

The rate is 8.5%, and the time is $\frac{1}{12}$ year.

$i = \$65,000 \times 0.085 \times \frac{1}{12} = \mathbf{\$460.42}$

EXAMPLE 2 How much of the first month's payment is for the principal?

Solution Subtract the interest from the monthly payment.

$\$564.09 - \$460.42 = \mathbf{\$103.67}$

EXAMPLE 3 For the second month, the interest is calculated on what new principal?

Solution Subtract the first month's principal payment from the original principal.

$\$65,000 - \$103.67 = \mathbf{\$64,896.33}$

The amount of the second month's payment that goes toward interest is calculated on the new principal, $64,896.33.

EXAMPLE 4 What total amount will Phil and Patty pay if they make regular monthly payments for the full 20 years of the mortgage?

Solution 20 years = 20 × 12 = 240 months

$240 \times \$564.09 = \mathbf{\$135,381.60}$

To solve the problems in the next exercise, review:
- finding a percent of a number, page 236

Part A

Use the following table to answer problems 1 to 5.

Approximate Monthly Mortgage Payment (30-year fixed rate)				
Amount $	6% $	7% $	8% $	9% $
50,000	307	333	367	402
75,000	450	499	550	603
100,000	599	665	734	804
125,000	749	831	917	1,006
150,000	899	998	1,100	1,206
175,000	1,049	1,164	1,284	1,408
200,000	1,199	1,330	1,468	1,609

1. On a 30-year mortgage for $100,000 at 8%, what is the monthly payment?

2. What is the monthly payment on a $75,000 mortgage at 9%?

3. Heather doesn't want to spend more than $1,000 per month on her mortgage payments. At a 7% annual interest rate, what is the largest mortgage shown in the table that she can afford?

4. For a $50,000 mortgage, the monthly payment at 9% is how much more than the monthly payment at 6%?

5. Kathy and Matt can pay $1,200 a month on a mortgage. At 6% annual interest, what is the largest mortgage amount that they can afford according to the table?

Part B

Use the following information to answer problems 6 to 10.

SITUATION

To buy a condominium, Saori got a 15-year, fixed-rate mortgage for $98,000. The interest rate is 5.625%, and her monthly payment is $860.44.

6. How much of her first month's payment is for interest?

7. How much of her first month's payment goes toward paying off the principal?

8. The interest for the second month's payment is based on what new principal?

9. How much of the second month's payment is for interest?

10. What amount of the second month's payment is for principal?

Part C

Use the following table to answer problems 11 to 15.

Approximate Monthly Mortgage Payment (8.5% fixed rate)				
Amount $	10-yr $	15-yr $	20-yr $	30-yr $
50,000	620	492	434	384
75,000	930	739	651	577
100,000	1,240	985	868	769
125,000	1,550	1,231	1,085	961
150,000	1,860	1,477	1,302	1,153
175,000	2,170	1,724	1,519	1,346
200,000	2,480	1,970	1,736	1,538

11. What is the monthly payment on a $75,000 mortgage at 8.5% for 30 years?

12. Find the monthly payment on a $150,000 mortgage at 8.5% for 20 years.

13. For a $200,000 mortgage at 8.5% for 10 years, what is the monthly payment?

14. Marcus and Blanca cannot spend more than $1,000 on their mortgage payment. For the mortgages listed in the table, what is the largest amount they can borrow for 30 years at 8.5%?

15. Mia got a 10-year mortgage for $50,000 at 8.5%. How much will she pay all together if she pays off her mortgage in regular monthly installments?

Part D

Use the following information to answer problems 16 to 18.

SITUATION

Roberto and Jeanne got a 30-year, fixed-rate mortgage for $122,000. The interest rate is 6.75%, and their monthly payment is $791.30.

16. How much of their first month's payment is for interest?

17. How much of their first month's payment is for paying off the principal?

18. The interest for the second month's payment is based on what new principal?

Post-Lesson Vocabulary Reinforcement

Match the definitions with the terms. Write the appropriate letters in the blanks.

a. the property itself

b. a mortgage

c. a usual down payment on a mortgage

d. an adjustable rate of interest

e. the cap

f. a common long-term

g. a fixed rate of interest

_____ 1. an installment loan for the purchase of a house, condominium, or apartment

_____ 2. the collateral for a mortgage

_____ 3. at least 20% of the asking price of a property

_____ 4. the interest rate that never changes for term of mortgage

_____ 5. the interest rate that changes every year

_____ 6. the maximum or upper limit the interest rate can rise

_____ 7. a 30-year mortgage

In the blanks, write _T_ if the statement is true and _F_ if the statement is false.

_____ 8. A condominium could be the collateral for a mortgage.

_____ 9. The amount that an ARM can change each year is limited to four percentage points.

_____ 10. A long-term mortgage has the advantage of lower monthly payments.

_____ 11. The _cap_ is another word for the minimum or highest amount.

Now compare your answers with a partner and then correct any statements that are false. Next compare your corrections and your first answers with the class.

Fill in the blanks using words from this lesson. Write one letter on each blank line.

12. An installment loan for the purchase of a house is called
    a __ __ __ __ __ __ __ __.

13. The upper limit to which the interest rate can rise is called
    the __ __ __.

14. The interest rate never __ __ __ __ __ __ __ for a fixed-rate mortgage.

15. For an adjustable-rate mortgage (ARM), the interest rate changes
    every __ __ __ __.

Language Builder

Note that -*self* as in "The collateral for a mortgage is the property its*self*" indicates the reflexive form of the pronoun. Its use here is simply to emphasize the noun, *property*.

ACTIVITY A In this lesson there are examples of **phrasal verbs** or **two-word verbs.** Here's an example: *The borrower pays off the debt more quickly.* Phrasal verbs are actually a verb + a preposition, in this case *pays* + *off.* These two-word verbs, as they are sometimes called, have their own meanings—for example, as you learned earlier, *pay off* means more than simply *pay.* See if you can identify more two-word verbs in the following sentences, adapted from the lesson. Underline them.

1. The buyer has paid off the mortgage.

2. The buyer puts down at least 20% of the asking price.

3. This amount goes toward interest on the new principal.

ACTIVITY B Write the correct two-word verbs from the box in the blanks. Make sure to use the correct tenses and endings of the verbs. To check your work, look for similar sentences in the lesson.

put down	go toward	pay off	paid off

4. Yesterday Phil and Patty finally _____ their mortgage.

5. Mortgage lenders like prospective buyers to _____ at least 20% of the asking price of a property.

6. The amount of the second month's payment that _____ interest is calculated on the new principal.

7. How much will Mia pay all together if she _____ her mortgage in regular monthly installments?

ACTIVITY C From previous lessons, you learned that a **superlative adjective** (a word that may end in -*est*) means that the noun that follows it is the maximum or minimum. If there are superlative adjectives in the following sentences, underline them. If not, don't underline anything.

8. At a 7% annual interest rate, what is the largest mortgage shown that Heather can afford?

9. The largest amount that Kathy and Matt can afford is $1,200 a month.

10. What is the least amount Saori has to pay for her first month's interest payment?

UNIT 3
UNAVOIDABLE
EXPENSES

Lessons 14 to 19 cover some unavoidable financial situations, including paying utility bills, renting or buying a home, buying insurance, and paying taxes. Again, many of these topics have a specialized vocabulary. Learn any new terms as well as the mathematical procedures involved in calculating these expenses.

LESSON 14: UTILITIES

Pre-Lesson Vocabulary Practice

Study the following terms and their meanings. Then find them in the lesson and read the surrounding sentences to see if you better understand the meanings. Finally, take turns with a partner choosing a term from the list and then giving the meaning.

natural gas – a substance used as energy for cooking, heating, etc.

commodity – a product or service

are subject to – must obey/conform to/follow

regulations – rules

occupants – persons who live in a particular place

to consume – to use

corresponds to – represents or signifies

alternate between – go from one to another

provider – company that supplies a product or service

cubic foot – a volume having six equal sides of one square foot each

to complicate – to make (something) difficult

flat fee – pre-determined or fixed amount

voice mail – service that records incoming calls for the customer to listen to later

call waiting – the telephone service that allows the user to make a caller wait while the user answers another incoming call

call forwarding – the telephone service that allows the user to have incoming calls forwarded or sent on to another number

caller ID – (caller identification) the telephone service that allows the user to know who is calling when the telephone rings

interstate telephone network – a system of telephone service connecting two or more states

switch – change

affordable – not too expensive

rural – the opposite of city or urban

hold down the cost – keep the cost from going up

hearing impairment – a difficulty or problem with being able to hear

forms of fraud – kinds of illegal activity in which victims are deceived

Utilities

A **utility** is something useful such as electricity, water, natural gas, or telephone service. A **public utility** is a private business organization that provides a **commodity** (like water or natural gas) or a **service** (like telephone) to the public. All public utilities are subject to government regulations.

Consumers receive regular bills for the utilities they use. Electricity, gas, and telephone bills arrive each month. Water and sewer bills are often sent quarterly. The mathematics of most utility bills is simple. A bill reports **usage**—the amount of the utility that a customer used during the time period of the bill—and charges. To calculate the basic cost of a utility, a company uses the formula $c = nr$ where c stands for cost, n stands for the number of units that a consumer uses, and r is the rate or cost for one unit of the utility. The total bill may include other fees such as delivery charges and taxes.

The unit of measurement for electricity is the **kilowatt-hour** (kwh). One kilowatt-hour of electricity will light a 100-watt light bulb for 10 hours. Every house or apartment has a meter that measures the number of kilowatt-hours that the occupants consume. Most electric meters have five dials. Each of the dials has ten digits (0 to 9) arranged in a circle.

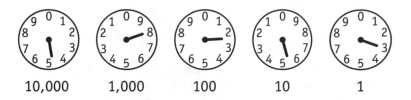

10,000   1,000   100   10   1

KILOWATT–HOURS

Below the dials on an electric meter are the numbers 10,000, 1,000, 100, 10, and 1. These numbers correspond to the first five places in our number system. Notice that the digits on the dials alternate between a clockwise arrangement and a counter-clockwise arrangement.

When the hand on a dial is between two digits, the reading is the smaller digit. For example, if a hand on a dial is between 5 and 6, the reading is 5.

For the five dials in the illustration above, the reading is 48253. The number represents 48,253 kilowatt-hours of electricity.

EXAMPLE 1 The dials in the illustration above represent a reading of the electric meter for Carla's apartment. If the meter reading one month ago was 48089, how many kilowatt-hours of electricity did Carla use during the month?

Solution Subtract the previous reading from the current reading.

$48,253 - 48,089 =$ **164 kilowatt-hours**

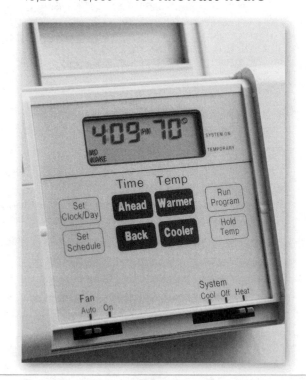

EXAMPLE 2 Carla's electricity provider charges 16.7612¢ per kilowatt-hour. Find the total cost of the electricity Carla used for the month.

Solution Multiply the number of kilowatt-hours of electricity that Carla used by the rate. The rate of 16.7612¢ is $0.167612.

$c = nr = 164 \times \$0.167612 = \$27.488\ldots$ or **$27.49**

Water and natural gas are usually measured in cubic feet and are billed in units of 100 cubic feet. The abbreviation for 100 cubic feet is either *hcf* or *ccf*. The first *c* in *ccf* stands for the Roman numeral *C*, the symbol for 100. To complicate things, 100 cubic feet of gas is often referred to as one **therm.**

EXAMPLE 3 The current reading on Mrs. Adams' gas meter is 3956 ccf. The reading last month was 3924 ccf. The company that provides natural gas to her area charges $0.4465 per ccf. What is the monthly charge for the gas that Mrs. Adams used for the month?

Solution Subtract the previous meter reading from the current reading.

$3{,}956 - 3{,}924 = 32$ ccf

Multiply the number of units used by the rate per unit.

$32 \times \$0.4465 = \$14.288$ or **$14.29**

On natural gas bills there may be a **basic monthly service fee** as well as the usage charge and state and local taxes.

Water is also measured in cubic feet (cf) or 100 cubic feet (ccf or hcf). One cubic foot of water is about 748 gallons.

EXAMPLE 4 Ellen and Tito's community charges a basic monthly service fee of $11.36 for water. Customers pay an additional $1.395 per hcf. Find the monthly charge to Ellen and Tito if they use 16 hcf of water.

Solution   Find the cost of 16 hcf of water.

16 × $1.395 = $22.32

Add the basic service fee to the water charge.

$11.36 + $22.32 = **$33.68**

On water bills there is often a charge for sewers. The sewer charge may be a flat fee or it may be a percent of the amount of water the customer used.

Bills for **telephone service** do *not* resemble other utility bills. In recent years the cost of telephone service has become more and more complicated. The simple application of the cost formula has nearly disappeared from the modern phone bill. Telephone bills now include charges for local telephone service; for long-distance service, which is often provided by a company other than the local provider; for additional services such as voice mail, call waiting, call forwarding, and caller ID; and for federal, state, and local taxes.

Below is a list of some of the charges that commonly appear on telephone bills:

Subscriber Line Charge, also called the **Federal Access Charge** or the **FCC Line Charge.** This is a charge required by the Federal Communications Commission (FCC), the government agency that regulates telephone companies. This fee covers part of the cost for maintaining lines, wires, poles, equipment, and facilities that connect a home to the interstate telephone network.

Local Number Portability or **LNP.** This is a fee that allows customers to keep their number if they switch from one company to another.

Universal Service Charge or **Universal Connectivity Fee.** This is a charge that helps make telephone service affordable to rural and low-income consumers. It also helps hold down the cost of providing Internet access to schools, libraries, and rural health centers.

911 Service Fee. This is a fee for emergency systems that can locate a caller's address.

Telecommunications Relay. This is a charge for funding the special telephone system that allows a customer with a hearing impairment to make and receive calls.

The total amount due on a monthly telephone bill is the sum of many charges.

EXAMPLE 5 What is the total of the following charges for local telephone service?

Local Calling Plan	$18.00
Charge for Network Access	$ 6.50
Universal Connectivity Fee	$ 0.50
Local Number Portability	$ 0.30
Emergency 911	$ 1.00
Telecommunications Relay	$ 0.90
Federal Tax	$ 2.25
State Tax	$ 2.50
Local Tax	$ 0.50

Solution   Add the 9 items in the list. The total is **$32.45.**

It is often easier to simply pay a telephone bill than to look at all the detail on the bill. However, customers should read telephone bills thoroughly to be sure that they have been charged correctly.

There are frequent **scams,** or forms of fraud, with telephone bills. **Slamming** occurs when a long-distance carrier is changed without a customer's permission. **Cramming** occurs when a company charges for services that a customer never ordered. A customer who does not understand a charge on a telephone bill should call the company that sent the bill and ask for an explanation.

To solve the problems in the next exercise, review:

- adding decimals, page 230
- multiplying decimals, page 231

Part A

You may use a calculator to solve the problems.

1. What is the reading on the dials pictured below?

10,000   1,000   100   10   1

KILOWATT–HOURS

2. The dials pictured above represent the reading at the end of the most recent billing period for the electric meter on Miguel and Alisandra's house. At the cost of 13.462¢ per kilowatt-hour, what is the cost of the month's electricity if the reading for the previous month was 07605 kwh?

3. What is the reading on the electric meter dials pictured below?

10,000   1,000   100   10   1

KILOWATT–HOURS

4. The dials above represent the most recent reading on Jan's electric meter. What is the cost of electricity for the month if the reading last month was 23989 and the cost per kilowatt-hour is 11.3085¢?

5. The illustration below represents the dials on Tony's electric meter at the most recent reading. What is the reading in kilowatt-hours?

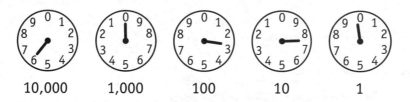

10,000   1,000   100   10   1

KILOWATT–HOURS

6. The cost per kilowatt-hour for Tony's electricity is 16.7612¢. What was the cost for Tony's electricity for the month if the previous meter reading was 60096 kwh?

Part B

The following electric bills are for Keith's home and for his woodworking shop next door. Use the bills to answer problems 7 to 13.

NYSEG

Account number:
79-122-07-034800-01

Your electric cost (Residential service)
You used 137 kwh at a cost of $26.97

Billing period: 30 days
06/17/03 to 07/17/03

Amount of electricity used
Latest reading 07/17 (NYSEG read)	35078—
Previous reading 06/17 (calculated)	34941
Electricity used (kwh)	137

Meter# 80136469

NYSEG Fixed Price

Cost of electricity used
Delivery & supply cost
Basic service charge (for 1 month)	$12.30
All 137 kwh @ 10.7076 ¢/kwh	+14.67
Electric cost	$26.97

NYSEG

Account number:
79-122-07-034700-02

Electric cost (Non-residential rate)

Billing period: 62 days
05/16/03 to 07/17/03

Amount of electricity used
Latest reading 07/17 (NYSEG read)	02786—
Previous reading 05/16 (NYSEG read)	02258
Electricity used (kwh)	528

Meter# 63996470

NYSEG Fixed Price

Cost of electricity used
Delivery & supply cost
Basic service charge (for 2 months)	$27.64
All 528 kwh @ 12.5109 ¢/kwh	+66.05
Sales tax (8.25%)	+ 7.73
Electric cost	$101.42

7. What is the length of the billing period for the electricity bill for Keith's residence?

8. What is the length of the billing period for the electricity bill for Keith's shop?

9. How many kilowatt-hours of electricity did Keith use in his home for the time period shown on the bill?

10. How many kilowatt-hours of electricity did Keith use in his shop for the time period shown on the bill?

11. What is the basic monthly service charge for electricity to Keith's home?

12. What is the *average* basic monthly service charge for electricity to Keith's shop?

13. If next month's reading for Keith's home is 35231, calculate his home electric bill for the month.

Part C

Use the following information to answer problems 14 to 17.

SITUATION

The company that provides natural gas to Carol and Nicholas's house includes the following charges each month:

A. a charge of $.266 per ccf of gas
B. a distribution charge of $.132 per ccf of gas
C. a regular customer fee of $7.00
D. state tax of 4%

14. The gas meter at Carol and Nicholas's house has a reading for the current month of 4168 ccf. The reading for the previous month was 4098. How many hundred cubic feet of natural gas did they use?

15. What was the basic charge for gas for the month?

16. What was the distribution charge for the month?

17. What was their total natural gas bill?

Part D

Mrs. Anderson's municipal water supply company includes the following charges each month. Use this information to answer problems 18 and 19.

Water base fee	$11.75
Water usage	$1.395 per hcf
Sewer base fee	$10.25

18. Mrs. Anderson's water meter had a reading of 931 hcf this month and 917 hcf last month. How much water did she use during the month?

19. What was Mrs. Anderson's total bill for water and sewer for the month?

Part E

Use the following bill for Jake's local telephone service to answer problems 20 and 21.

Local Services

Basic service and calls

Local service	$19.95
FCC line charge	6.44
Local number portability	0.23
Universal connectivity	0.59

Custom calling features

Call blocking	$3.00
Voice mail	6.00

Taxes and surcharges

Federal tax	$0.86
911 surcharge	1.00
Local tax	1.26
State tax	1.23

20. What is the total for the local portion of Jake's telephone bill?

21. Which items on the bill can Jake cancel or question if he cannot remember ordering them?

Post-Lesson Vocabulary Reinforcement

Choose terms from the list on the right to complete the sentences below.

1. The numbers 10,000, 1,000, 100, 10, and 1 correspond to _____ in our number system.

2. The _____ for 100 cubic feet is either hcf or ccf.

3. The government agency (FCC) that regulates telephone companies is called the _____.

4. A _____ is something useful such as electricity, water, natural gas, or telephone service.

5. Subscriber Line Charge is also called the Federal Access Charge or the _____.

6. One hundred cubic feet of gas is often referred to as one _____.

7. The first *c* in *ccf* stands for the Roman numeral *C*, _____.

8. _____ occurs when a company charges for services that a customer never ordered.

9. _____ is a fee that allows customers to keep their number if they switch from one company to another.

10. _____ is a fee for emergency systems that can locate a caller's address.

11. There are frequent _____, or forms of fraud, with telephone bills.

12. _____ is a charge for funding the special telephone system that allows a customer with a hearing impairment to make and receive calls.

13. _____ occurs when a long-distance carrier is changed without a customer's permission.

14. A _____ is a private business organization that provides a commodity to the public.

15. The unit of measurement for electricity is the _____ (kwh).

16. _____ is the amount of a utility that a customer used during the time period of the bill.

17. An example of a service to the public is _____ service.

18. Examples of commodities are _____.

utility

the first five places

Local Number Portability or LNP

abbreviation

the symbol for 100

Federal Communications Commission

cramming

FCC line charge

therm

telephone

scams

usage

kilowatt-hour

public utility

slamming

911 service fee

water and natural gas

telecommunications relay

Language Builder

ACTIVITY A You have already learned that **clauses** are sentences within a sentence and that many clauses begin with clause markers. To review clauses and clause markers, put the words in order on a separate sheet of paper.

1. A public utility is a private business organization (a commodity   that   provides).

that provides a commodity
_____

2. Every house has a meter (measures   the number   that   of kilowatt-hours).

3. This is a fee (to keep   allows   that   customers   their telephone number).

4. (between two digits   When   is   the hand on a dial), the reading is the smaller digit.

ACTIVITY B Now practice completing the following sentences with your own clauses. Then compare your work with a partner's and with the class.

5. The utility that _____ is my telephone service.

6. I know someone who _____.

7. When I _____, I am very happy.

8. There are some utilities that _____.

9. Do you know anyone who _____?

ACTIVITY C Remember **modals**—words that are used with verbs to express a particular meaning? You've already learned about the modals *can* and *will*. Two new modals from this lesson are *should* and *cannot*—the negative of *can*. The modal *should* expresses "obligation"—that it's necessary to do something. Underline the modals in the following sentences from this lesson.

_____ **10.** Customers should read telephone bills thoroughly to be sure that they have been charged correctly.

_____ **11.** One kilowatt-hour of electricity will light a 100-watt light bulb for 10 hours.

_____ **12.** Which items on the bill can Jake cancel or question if he cannot remember ordering them?

Now match the following meanings with the sentences above. Write the letter of the meaning in the appropriate blank above.

a. a future or emphatic action   **b.** a possibility

c. a negative possibility   **d.** an obligation

Pre-Lesson Vocabulary Practice

Study the terms on the right and then find them in the lesson. Read the surrounding sentences to see if you better understand the meanings. Finally, use the terms to fill in the blanks below. Compare your answers with a partner.

1. Every family has a _____ who is in charge of the family.

2. The branch of the U.S. government responsible for counting and reporting information on how many people live in the U.S.A. is the _____.

3. To put one's signature on a legal rental document is to _____.

4. Homes that can be moved from one place to another are called _____.

5. Some people live in boats that have living quarters called _____.

6. _____ are temporary guest residences.

7. Properties such as apartments that are lived in but not owned are _____.

8. A _____ is the person who owns property but rents it for a charge.

9. A _____ is the person who rents property.

10. When you sign a lease, you have a _____ to pay the rent.

11. Houses usually _____ in value.

12. Cars don't gain value; instead they _____.

13. _____ is charged for living in a house or apartment that one does not own.

14. A renter is often required to leave money called a _____ as a guarantee.

15 Real estate _____ help people find a place to live.

16 Another name for a payment is an _____.

17. Attorneys, sometimes called _____, should be hired to assist in the purchase of a house.

head of household

hotel/motel rooms

houseboats

landlord

rental properties

sign a lease

trailers/mobile homes

U.S. Census Bureau

agents

appreciate

depreciate

expenditure

financial responsibility

lawyers

rent

security deposit

tenant

Renting or Buying a Home

Housing includes apartments, private houses, trailers, houseboats, and hotel/motel rooms. The **head of a household** has two choices when deciding how to pay for housing: to rent or to buy. According to the U.S. Census Bureau, about 68% of the homes in America are occupied by owners or by consumers who are paying mortgages.

The owners of rental properties usually ask renters to sign a **lease.** A lease is a contract, or legal agreement, between a **landlord** and a **tenant.** A lease specifies the monthly rent and the length of time of the agreement (often one or two years). A lease may also include certain restrictions such as limitations on late night noise and the keeping of pets. In most places landlords are free to charge what the market is willing to pay.

At the starting time of a lease, a renter may have to pay an extra month's rent as a **security deposit.** The tenant may also have to pay a fee to the **agent** who showed the apartment or the house. After paying the initial charges, the renter's only financial responsibility is to pay the monthly rent. If there is a plumbing problem, the owner has to take care of it.

EXAMPLE 1 Maggie and Charles rented a two-bedroom apartment for $540 a month. The landlord required an extra month's rent as a security deposit, and the renting agent required a fee of 80% of the first month's rent as a finder's fee. How much did Maggie and Charles have to pay in rent and fees for the first month?

Solution The rental agent's fee is 80% of $540.

$0.8 \times \$540 = \$432$

Add to find how much Maggie and Charles had to pay.

$\$540 + \$540 + \$432 =$ **$1,512 the first month**

EXAMPLE 2   After a year in their apartment, Maggie and Charles decided to renew their lease. The new rent will increase by 6.5%. What will the new monthly rent be?

Solution   **Find 6.5% of $540.**

$0.065 \times \$540 = \$35.10$

The new rent will be

$\$540 + \$35.10 =$ **$575.10 per month**

Buying a home requires far more cash up front than renting. As you learned in the lesson about mortgages, most home buyers do not pay the total purchase price of their house or apartment in cash. Instead, they make a **down payment** and borrow the balance from a lending agency such as a bank.

Mortgage lenders like buyers to **put down** at least 20% of the asking price of a home.

EXAMPLE 3   Sal and Nora want to buy a house that is listed for $165,000. How much will they have to borrow if they make a down payment of 20% of the asking price?

Solution   **Find 20% of $165,000.**

$0.2 \times \$165,000 = \$33,000$

Subtract the down payment from the asking price.

Mortgage amount $= \$165,000 - \$33,000 =$ **$132,000**

One advantage to buying rather than renting is that the value of a house usually increases over time. A house or condominium is an **appreciating asset.** An asset is anything of value that someone owns. *To appreciate* means to "increase in value." This is the opposite of a car, which **depreciates** in value as soon as the owner drives away from the car dealer's lot.

EXAMPLE 4 Bill and Tina bought their house in 1995 for $109,000. They sold the house in 2002 for $148,000. By what percent did the market value of the house appreciate?

Solution Find the difference between the 2002 price and the 1995 price.

$148,000 − $109,000 = $39,000

Divide the increase in the price by the original price.

$$\frac{\$39,000}{\$109,000} = .357\ldots \text{ or } \textbf{about 36\%}$$

The down payment for a home is the biggest expenditure for most buyers, but there are other expenses as well. When a seller and a buyer agree to a price and the buyer has arranged for **financing,** the two parties and their lawyers select a **closing date.** This is the day when the final settlement is made and legal documents are signed. Several **closing costs** have to be paid on this date.

Some of the costs in buying a house are:

Origination fee—This is a percent (often 1%) of the total mortgage amount that is charged by the lender.

Title insurance—This guarantees that the previous owner is, in fact, the owner.

Escrow fees—An escrow account is the place where money that has already been paid toward the purchase is kept.

Homeowner's insurance—This is required before a new owner can move into the house.

Appraisal fees—This is the expense of having an inspector look over a property to be sure of its value. The appraisal is often used to calculate new taxes.

Credit report fees—The mortgage lender requires a credit check on the buyer.

Property taxes—If a seller has paid taxes for part of the time that a new buyer will occupy a property, the buyer owes that portion of the taxes to the seller.

State recording fees—This varies from state to state but is usually $40 to $80.

Notary public charge—This assures that the seller is the person he claims to be.

Legal fees—Most buyers use lawyers to write and assemble the documents required for a property purchase.

EXAMPLE 5   Mr. and Mrs. Miller have decided to buy a condominium for $130,000. They plan to make a down payment of 20%. Their lawyer told them that their total closing costs, including the lawyer's fees, will be $1,950. How much cash will they have to spend to buy their condominium?

Solution   Find 20% of $130,000.

$0.2 \times \$130,000 = \$26,000$

Their total cash expenditure is the down payment plus the closing costs.

$\$26,000 + \$1,950 = \mathbf{\$27,950}$

Every homeowner must also remember that expenses for repairs will occur from time to time.

To solve the problems in the next exercise, review:
- adding decimals, page 230
- finding a percent of a number, page 236
- finding what percent one number is of another, page 238

EXERCISE 15

Part A

Use a calculator to solve the problems.

1. Fred signed a lease for an apartment that costs $490 a month. He has to pay the first month's rent, the last month's rent, and a security deposit of $300. How much does he have to pay all together when he first moves into the apartment?

2. After a few months in his apartment, Fred realized that the additional expenses for utilities were about $65 every month. Not including the security deposit, approximately what total amount will Fred spend in a year for housing?

3. Maya, Victoria, and Evelyn decided to share equally the costs of a house that rents for $810 a month. The landlord asked them to pay two months' rent at the beginning of their lease plus a deposit of $750. How much did each woman have to pay the first month?

4. How much will each of the women in problem 3 have to pay in rent each month?

5. Carmen and Jesse now pay $604.70 a month for their two-bedroom apartment. At the end of their first year in the apartment, the landlord offered them a two-year renewal with a rent increase of 4.5%. What will be the new monthly rent?

6. A guideline for paying rent is not to spend more than 30% of one's gross income. Charlene makes $31,400 a year. Using the 30% guideline, what is the maximum monthly rent that she can afford?

Part B

Use the following information to answer problems 7 to 11.

SITUATION

> Sandy and Doug want to buy a condominium apartment that costs $129,000. Their gross income is $28,500 from Doug's job as a security guard and $6,500 from Sandy's part-time teaching job. By combining their savings and gifts from their parents, Sandy and Doug are able to make a down payment of 24% of the asking price of the apartment.

7. What is the couple's total yearly income?

8. What is the down payment they will make?

9. How big is the mortgage that they will have to take out?

10. Sandy and Doug's lawyer told them to expect to pay an additional $2,350 in expenses at the closing. How much do they have to pay all together in cash at closing to buy the condominium?

11. The lending officer at the bank where they got their mortgage said that they should not spend more than 30% of their gross income on mortgage payments. The monthly payment on their 15-year mortgage at 6.25% is $797.42. Do Doug and Sandy have enough income to cover their monthly mortgage payments?

Part C

The table below shows the median price of existing houses in eight metropolitan areas for a recent year. Use the table to answer problems 12 to 16.

Little Rock, Arkansas	$129,800
Albany, New York	$197,900
Los Angeles, California	$402,100
Dayton, Ohio	$107,000
Springfield, Illinois	$108,000
Eugene, Oregon	$224,700
Cedar Rapids, Iowa	$136,500
San Antonio, Texas	$152,800

12. For which city on the list is the median price of a house highest?

13. For which city on the list is the median price of a house least expensive?

14. How much cash is required for a 20% down payment on a house at the median price in Eugene, Oregon?

15. How much more is the median price of a house in Los Angeles than the median price of a house in Albany?

16. The median price of a house in Springfield, Illinois, is what percent of the median price of a house in Cedar Rapids, Iowa?

Part D

The following closing statement describes the terms of the sale of a house to Mr. and Mrs. Cartwright. Use the statement to answer problems 17 to 20.

STATEMENT OF SALE

Dated: February 20, 2003

Mr. & Mrs. Smith _____ to _Mr. & Mrs. Cartwright_____

Street Address: 20 Maple Lane_____

Town: Pleasant Hill   **County:** Central   **State:** New York

Purchase Price	$145,000.00
Insurance Adjustment	$
Fuel Oil Adjustment	$
Village Taxes (Adjusted)	$ 59.36
County Taxes (Adjusted)	$ 184.64
School Taxes (Adjusted)	$ 114.19
Total Amount Due Seller	$145,358.19

Credits to Purchaser:

Amount Paid Down	$ 14,500.00	
First Mortgage	$	
Water and Sewer	$ 119.20	
		$ 14,619.20

Balance Due	$130,738.99
Mortgage Proceeds	$104,000.00
Cashier's Check	$ 26,738.99

Expenses of Purchaser:

Title Insurance	$ 573.00
Lender's Attorney's Fees	$ 325.00
Recording Fees	$ 52.00
Discount Fee	$ 1,040.00
Escrow	$ 414.12
Purchaser's Attorney's Fees	$ 700.00
Total	$ 3,104.12

---

17. The cash that applies to the purchase price of the house is the amount paid down plus the amount of the cashier's check that the Cartwrights brought to the closing. Find the total amount of cash that they paid to purchase the house.

18. The cash that the Cartwrights paid was what percent of the total purchase price of the house?

19. Mr. and Mrs. Smith had to pay the real estate agent who sold their house a commission of 6% of the sale price. What was the agent's commission?

20. Mr. and Mrs. Cartwright got a 30-year mortgage for $104,000 at 7%. The monthly payment is $691.93. The Cartwrights' yearly income is $32,300. What percent of their yearly income are their mortgage payments?

Post-Lesson Vocabulary Reinforcement

Match the definitions with the terms on the right. Write the appropriate letters in the blanks.

_____ **1.** a contract, or legal agreement, between a landlord and a tenant

_____ **2.** anything of value that someone owns

_____ **3.** money charged by a lawyer for writing and assembling the documents required for a property purchase

_____ **4.** the day when the final settlement over the purchase of a house is made and legal documents are signed

_____ **5.** varies from state to state, but is usually $40 to $80

_____ **6.** money paid by the buyer to have a credit check done

_____ **7.** the expense of having an inspector look over a property to appraise its value

_____ **8.** a special account where money is placed that has already been paid toward the purchase of a piece of property such as a house

_____ **9.** a guarantee that the previous owner is, in fact, the owner of a piece of property

_____ **10.** money—often 1% of the total mortgage amount—that is charged by the lender

a. a lease

b. state recording fees

c. credit report fees

d. closing date

e. appraisal fees

f. an asset

g. escrow

h. title insurance

i. origination fee

j. legal fees

Think back to the Language Builder from Lesson 7. You were introduced to W/H information questions. Go back and review them if needed. Use the information from this lesson to answer the following questions. As a sample, the first one has been done for you.

11. What two choices does the head of a household have when deciding how to pay for housing? _to rent or to buy_

12. If there's a plumbing problem for a renter, who has to take care of it?

13. What is a renter's only financial responsibility after paying initial charges?

14. Why do some items increase or appreciate in value?

15. What is an example of something that depreciates in value?

16. When do a seller and a buyer and their lawyers select a closing date?

17. As part of the closing process, where is money placed that has already been paid toward the purchase of a piece of property such as a house?

Language Builder

ACTIVITY A In Lesson 2, you learned about **gerunds**—nouns that end in *-ing*. Here is some additional information about gerunds and some practice using them.

Quickly read through the main part of this lesson and look for gerunds. Start with the title, **Renting or Buying a Home,** which has two gerunds. Now <u>underline</u> the gerunds in the following sentences from this lesson. Be careful not to underline present participles or other *-ing* words.

1. Housing includes apartments, private houses, trailers, houseboats, and hotel rooms.

2. The head of a household has two choices when deciding how to pay for housing: to rent or to buy.

3. According to the U.S. Census Bureau, about 68% of the homes in America are occupied by owners or by consumers who are paying off their mortgages.

4. After paying the initial charges, the renter's only financial responsibility is to pay the monthly rent.

5. Buying a home requires far more cash up front than renting.

In Lesson 2, you also learned about **infinitives** (to + simple form of the verb). Now double underline the infinitives in the exercise above.

ACTIVITY B Sometimes we can use either a **gerund** or an **infinitive.** That's usually when they both work in the same way in a sentence—as nouns. If you think that's possible with the following sentences, rewrite the sentences on a separate sheet of paper using the other form.

6. Renting or buying a home is an important decision.

7. The head of a household has two choices when deciding how to pay for housing.

8. To rent a house doesn't require as much cash up front as to buy one does.

9. Buying a home requires far more cash up front than renting one.

10. After a year in their apartment, Maggie and Charles decided to renew their lease.

ACTIVITY C Use your own words to complete the following sentences with gerunds or infinitives.

11. I like _____ gerunds and infinitives in sentences.

12. _____ a house requires a lot of careful planning.

13. After _____ this exercise, I'm going _____.

14. If I plan _____ a house, I have to decide how _____ for it.

LESSON 16: CAR EXPENSES

Pre-Lesson Vocabulary Practice

Study the following terms and their meanings here and in the lesson itself. Next take turns with a partner choosing a term from the list and then giving the meaning. Finally, practice using the terms in sentences with a partner.

registered – officially counted and recorded

the automobile industry – the business of manufacturing and selling cars and other vehicles

keep a vehicle running/keep a car in good running order – maintain a vehicle so it works well

oil – the liquid used in a vehicle for lubrication

efficiency/efficient/efficiently – costing less money and giving better results

consumption – use

miles per gallon – miles driven for each gallon of gasoline used

filter – a part used in a vehicle to keep the oil clean

license plates – the tag that identifies a vehicle

leases – rents or borrows

worn – damaged from a lot of use

automobile club membership – to be a member of an association that helps automobile owners

rare/collectible – refers to older cars usually in good condition

resale value – what an item is worth when sold again

Study the pictures and related terms below. Next take turns with a partner: cover each term and point to the picture, and see if your partner can say the corresponding term.

sedan　　**pickup truck**　　**SUV**　　**two-door compact**

Car Expenses

The population of the U.S. is now over 300 million people. There are over 135 million registered automobiles in the U.S. This means that there is almost one car for every two people. In most communities the **public transportation** system is not adequate for moving people from their homes to their jobs, schools, shops, and other places people go. A car is often an expensive necessity.

The automobile industry offers many choices: two-door compacts, four-door sedans, station wagons, minivans, sport-utility vehicles (SUVs), and trucks. The prices and maintenance costs for vehicles differ widely.

To keep a vehicle running, an owner has to buy fuel—either **gasoline** or **diesel**—and change the oil from time to time. The price of fuel varies according to worldwide production levels. The amount of fuel that a car uses depends on both the number of miles driven and the efficiency of the car's engine. A **fuel-efficient** car uses relatively little gasoline. **Fuel** consumption is measured in **miles per gallon** (mpg). All cars use fuel more efficiently on highways than in cities, where drivers must start and stop frequently.

EXAMPLE 1 Celeste drove her new car 15,000 miles last year. Her records show that she bought a total of 525 gallons of gasoline during the year. To the nearest tenth, what average number of miles did Celeste drive on a gallon of gasoline?

Solution Divide the total number of miles she drove by the number of gallons of gasoline that she bought.

$15,000 \div 525 = 28.57\ldots$ or **28.6 mpg**

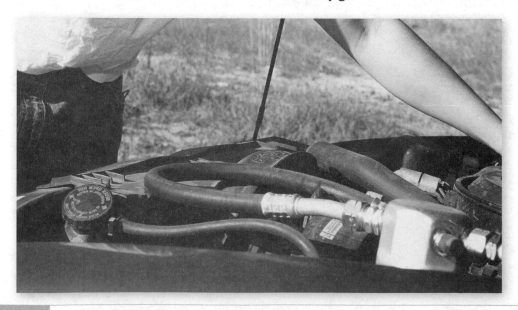

The label **mpg** (miles per gallon) is a clue to the solution to the last problem. The word *per* suggests division. To find the fuel efficiency of a vehicle, divide the number of miles by the number of gallons.

EXAMPLE 2   Celeste bought gasoline at an average price of $2.58 a gallon. How much did she spend on gasoline for the year?

Solution   Multiply the total number of gallons by the cost of one gallon of gasoline.

525 × $2.58 = **$1,354.50**

EXAMPLE 3   Celeste takes her car into the dealer for an oil and filter change every 2,500 miles. Her dealer charges $24.95 for this service. In the year when she drove 15,000 miles, how much did Celeste spend on oil changes?

Solution   Find how many times 2,500 divides into 15,000.

15,000 ÷ 2,500 = 6 times

Multiply the cost of servicing the car by 6.

6 × $24.95 = **$149.70**

Car owners have several expenses that have nothing to do with keeping their cars in good running order. A car owner has to pay a **registration fee,** a fee for **license plates,** and **insurance costs.** (You will learn more about insurance in Lesson 17.) If the owner borrowed money to buy the car, he or she has to make monthly payments on the loan. If the driver leases the car, he or she has to make monthly leasing payments.

EXAMPLE 4 Celeste took out a loan to buy her new car. The first year she spent $648 in interest on her loan. She also paid $1,150 for insurance and $935 in sales tax and registration fees. Find the total of these costs.

Solution   Add the costs.

$648 + $1,150 + $935 = $2,733

Celeste paid **$2,733.**

A car owner has to pay for regular maintenance, for occasional repairs, and for the replacement of worn parts. For new cars, the cost of maintenance and repairs is usually low. As a car gets older, these costs tend to rise. A car owner may also have to pay tolls for driving on highways, garage rent, parking lot fees, parking meter fees, automobile club membership fees, and occasional charges for cleaning.

One of the biggest expenses for car owners is **depreciation.** As you learned in the last lesson, houses normally increase in value over a period of years. A house is an *appreciating* asset. A car is the opposite. A car loses value as soon as it leaves a dealer's lot. A car is a *depreciating* asset. In other words, if an owner sells a new car a year after he or she bought it, he or she will have to sell it for much less than was originally paid.

EXAMPLE 5   The car that Celeste bought cost $15,090. Celeste took good care of her car, but one year after she purchased it, the car was worth only $10,940. By what percent did the value of Celeste's car depreciate in one year?

Solution   To find the amount of the first year's depreciation, subtract the two values.

$$\$15,090 - \$10,940 = \textbf{\$4,150}$$

Find the percent of change in the values.

$$\frac{\$4,150}{\$15,090} = 0.275\ldots \text{ or } \textbf{about 28\%}$$

Some cars depreciate faster than others. Except for a few old cars that are thought of as collectibles, most cars lose value. A car buyer should think about the cost of depreciation whenever she thinks about the cost of a new car.

Note: Gasoline prices rise and fall. The prices in this book may be quite different from current prices.

To solve the problems in the next exercise, review:

- adding, subtracting, multiplying, and dividing decimals, pages 230–231

- finding a percent of a number, page 236

- finding what percent one number is of another, page 238

EXERCISE 16

Use a calculator to solve the following problems.

Part A

1. Serena drove 920 miles on 33 gallons of gasoline. To the nearest tenth of a mile, how many miles did she get on one gallon of gasoline?

Part B

Use the following information to answer problems 2 to 4.

SITUATION

In one year Riley drove 12,164 miles. That year he bought 723 gallons of gasoline for which he paid $1,930.

2. What average price did Riley pay for a gallon of gasoline?

3. To the nearest tenth, what is the fuel efficiency of Riley's car? In other words, what average number of miles did he drive on one gallon of gasoline?

4. Riley got an oil change every 3,000 miles. Each oil change cost $22.95. How much did he spend on oil changes during the year?

Part C

Use the following information to answer problems 5 to 7.

SITUATION

Lucy drove 9,708 miles one year. She bought 266 gallons of gasoline and paid $726 for the gasoline.

5. What average price did Lucy pay for a gallon of gasoline?

6. To the nearest tenth, how many miles did Lucy drive on a gallon of gasoline?

7. Lucy got three oil changes during the year. On average, how many miles did she drive before getting an oil change?

Part D

Use the following information to answer problems 8 to 12.

SITUATION

> Larry's car gets 26 miles per gallon when he drives in the city
> and 34 miles per gallon when he drives on highways. Larry drove
> 10,973 miles last year. 80% of his driving was on highways.

8. How many miles did Larry drive on highways last year? Round your answer to the nearest mile.

9. How many miles did Larry drive in cities last year? Round your answer to the nearest mile.

10. How many gallons of gasoline did Larry use driving on highways? Round your answer to the nearest gallon.

11. How many gallons of gasoline did Larry use driving in cities? Round your answer to the nearest gallon.

12. At an average cost of $2.899 per gallon, how much did Larry spend on gasoline during the year?

Part E

Use the following information to answer problems 13 and 14.

SITUATION

> Guadalupe paid $5,999 for a used car. The chart below shows
> the amount that Guadalupe can expect the value of her car to
> depreciate each year.

First Year	Second Year	Third Year	Fourth Year	Fifth Year
$1,050	$900	$775	$650	$550

13. What total amount will the car depreciate in five years?

14. What will be the resale value of Guadalupe's car in five years?

Part F

Use the following information to answer problems 15 to 17.

SITUATION

José bought a new car for $18,699. The table below shows
the expenses José had for the first year.

interest charges on car loan	$ 780
sales tax and fees	$ 1,339
insurance	$ 1,190
fuel	$ 1,204
maintenance and repairs	$ 724

15. What was the total of the expenses for José's car the first year?

16. The value of José's car depreciated $8,800 the first year. The first year's depreciation was what percent of the sale price of José's car?

17. Including depreciation, what were the total costs for José's car the first year he owned it?

Part G

Use the following table to answer problems 18 and 19.

Description	Purchase Price	Sale Value after 1 Year
4-year-old sedan	$ 6,298	$ 5,173
new SUV	$30,699	$18,949

18. By what percent did the value of the sedan depreciate after one year?

19. By what percent did the value of the new SUV depreciate after one year?

Post-Lesson Vocabulary Reinforcement

Choose terms from the list on the right to complete the sentences that follow.

1. There are _____ 130 million registered automobiles in the U.S.

2. In most communities the _____ is not adequate for moving people around.

3. _____ is often an expensive necessity.

4. To keep a vehicle running, an owner has to buy _____—either gasoline or diesel.

5. The word *per* as in *per gallon* suggests _____.

6. A _____ car uses relatively little gasoline or fuel.

7. Car owners have several _____ that have nothing to do with keeping their cars in good running order—for example, registration fees.

8. Rare, old cars don't _____ as fast as others.

a car

more than

public transportation system

fuel

division

expenses

fuel-efficient

depreciate

Here's a vocabulary game. Use words from this lesson. Follow the steps.

Step One: Write one letter on each blank line below.

9. The U.S. population is now over 280 __ __ __ __ __ ☐ ☐ people.

10. *Shops* is another word for __ __ __ __ ☐ __ .

11. Drivers must stop and start __ ☐ __ __ __ __ __ __ __ __ .

12. ☐ __ __ __ or damaged parts on vehicles have to be repaired or replaced.

Step Two: Copy the letters from each box above:

13. _____ _____ _____ _____ _____

Step Three: Move the letters around until they spell another word from this lesson. Use it to complete this question:

14. Are you a car _____ ? Answer the question and share both of your answers with a classmate.

Language Builder

ACTIVITY A This lesson provides many examples of **irregular verbs.** Below is a chart contrasting them with **regular verbs** found in this lesson.

	PRESENT/ INFINITIVE	PAST	PAST PARTICIPLE
Regular Verbs	use	used	used
	depreciate	depreciated	depreciated
	borrow	borrowed	borrowed
Irregular Verbs	is/be	was/were	been
	have	had	had
	get	got	gotten

Notice how the past and past participle forms change for the irregular verbs in the bottom half of the chart. Listed below are other irregular verbs in this lesson. Follow the example in the first item and fill in the missing forms.

Present/Infinitive	Past	Past Participle
1. buy	bought	
2. drive	drove	
3.	paid	paid
4. spend		spent

Compare your work with a partner as the teacher goes over the answers. Next write your own sentences using the verb forms in this exercise. Compare your work with a partner and with the class.

ACTIVITY B In Lesson 14, you learned about **modals.** For example, you learned that the modal *should* expresses "obligation"—that it's necessary to do something. Stronger or greater obligation is expressed by the modal *must* and by the verb constructions *has to* and *have to.* Here is an example from this lesson: "A car owner *must* buy insurance for his car."

Now choose the correct words from the parentheses and write them in the blanks.

5. A car owner (has to/have to) _____ pay a fee for license plates.

6. Celeste (must/should) _____ think about the cost of depreciation whenever she thinks about the cost of a new car.

7. Celeste will (has to/have to) _____ sell her car for much less than she paid for it.

8. You (must/should) _____ put money in the parking meter if you want to park there.

LESSON 17: INSURANCE

Pre-Lesson Vocabulary Practice

Study the terms and their meanings throughout the lesson.
Next take turns with a partner choosing a term and then giving
the meaning. Finally, practice using the terms in sentences.

loss – the opposite of gain; not having money one expects to have
> Example: *I experienced a financial loss.*

out-of-pocket – direct, immediate costs
> Example: *Out-of-pocket expenses are not covered by insurance.*

comprehensive – including many aspects
> Example: *People should buy comprehensive insurance for a new car.*

uninsured motorist – a driver who has no insurance
> Example: *Uninsured motorists are often sued in court for damages.*

lawsuits – legal action taken against a person or business
> Example: *Insurance protects against lawsuits.*

disasters – bad events such as floods, hurricanes, tornados, fires
> Example: *Insurance does not cover all disasters.*

national health plan – government-provided healthcare
> Example: *There is no national health plan in the United States.*

resources – assets, a way to pay for something
> Example: *Medicaid is available for people without sufficient resources.*

group/individual health plans – health insurance for numbers of people/for one person
> Example: *HMOs are a type of group health plan.*

co-payment – the amount paid by the insured person
> Example: *Co-payments are extra payments in addition to insurance.*

disability – not being able to work due to illness or accident
> Example: *Some people receive disability assistance after an accident.*

skilled nursing – professional care given to someone disabled, sick, or elderly
> Example: *Insurance may provide skilled nursing and housekeeping care.*

severe financial hardship – great difficulty having enough money to pay for necessities
> Example: *Families may experience severe financial hardship after an accident.*

Insurance

Insurance is protection against financial loss. An insurance **policy** is an agreement between the consumer (the **insured** or the **policyholder**) who pays for the policy and the company (the **insurer**) that offers the insurance. An insurance policy is a detailed document that describes the **coverage** (the dollar amounts and limits) that the insurer will pay in case of a financial loss.

Lawyers call the insured the **first party,** and they call the insurer the **second party.** Insurance policies offer protection for the cost of an injury to a **third party.** A third party could be a carpenter working on a homeowner's roof or the driver of the other car in a crash. Insurance has many specialized terms and expressions. This lesson will explain the most common insurance terms.

Automobile insurance protects a policyholder from the costs that result from accidents, vandalism, or theft. **Homeowner's insurance** and **renter's insurance** protect against loss from fire, theft, wind, and water damage as well as the costs that result from injuries that occur on the property to a third party. **Health insurance** protects against the high costs of treating illnesses and injuries. **Life insurance** protects a family against the loss of income that results from the policyholder's death.

The price a customer pays for insurance is the **premium.** When a policyholder makes a **claim** (a request for payment), he often has to pay a **deductible.** A deductible is the amount that the policy owner must pay on a claim. For most insurance policies, higher deductibles mean lower premiums. When shopping for insurance, a costumer should consider what out-of-pocket expenses he or she can comfortably afford.

EXAMPLE 1 During a storm, a tree branch fell on the Suarez's house. The cost to repair the roof was $690. Their homeowner's policy has a property deductible of $250. How much will the insurer pay to repair the Suarez's roof?

Solution   Subtract the deductible from the cost of the repair.

$690 − $250 = **$440**

Automobile insurance offers several kinds of protection.

First-party coverage is for the policy owner and the passengers in his car.

Third-party coverage is for damage that a policy owner causes to another person or to another person's property.

Liability refers to a legal obligation. Liability insurance is for medical payments or for property damage to a third party. Most states require automobile owners to have liability insurance.

Collision refers to a crash. Collision insurance is for the policyholder's car. There is usually a deductible for collision insurance.

Comprehensive insurance covers the cost of damage or loss of the policy owner's car caused by something other than a collision. Comprehensive insurance includes damage or loss from fire, floods, earthquakes, and theft. There is usually a deductible for comprehensive insurance.

Uninsured or **underinsured motorist** protection covers costs to the policy owner from accidents caused by a driver with inadequate insurance.

Towing insurance covers the cost of moving the policy owner's car in case of an accident or a breakdown.

No-fault insurance pays for each driver's injuries regardless of who caused the accident.

The cost of automobile insurance varies according to the age of the driver, the age of the car, the driver's sex, the number of miles the driver normally drives in a year, the main purpose of most of the driving (work or leisure), and the driving record of the driver. In the last lesson you learned that the value of a car depreciates quickly. But depreciation does not mean that car insurance premiums drop as a car gets older.

One way to lower an insurance premium is to purchase more than one kind of coverage from the same company. Many insurance companies offer discounts if you buy car insurance, homeowner's insurance, and life insurance from them.

EXAMPLE 2 The yearly premium for Mike's automobile insurance is $762. Since he also has a homeowner's policy with the company, he gets a 7.5% discount on his car insurance premium. How much can he save on his car insurance?

Solution Find 7.5% of $762.

$0.075 \times \$762 = $ **$57.15**

Mike can save **$57.15.**

Homeowner's insurance covers the replacement cost of a house and its contents. Any homeowner with a mortgage is usually required by the lending institution to carry homeowner's insurance.

Homeowner's insurance includes liability insurance. This protects the owner against the costs associated with injury to anyone who is on the property. If a worker gets injured repairing a roof, the owner's insurance protects against lawsuits from the injured worker.

EXAMPLE 3 Mei-ling and David's homeowner's insurance premium is 0.65% of the value of their house. If their house is worth $122,000, what is their yearly insurance premium?

Solution Find 0.65% of $122,000.

$0.0065 \times \$122,000 = $ **$793.00**

Homeowner's insurance does not cover all disasters. To protect a home from loss by an earthquake or a flood, a homeowner usually has to purchase additional insurance coverage—called a **rider.**

Health insurance is controversial. The U.S. is one of the few developed countries with no national health plan. For people 65 and older, the federal government offers **Medicare,** which pays partial healthcare costs. For people without sufficient income or resources, there is **Medicaid.** Everyone else is dependent on group or individual health plans. By some estimates, 15% of Americans have no health coverage. For some age groups, this percentage is much higher.

Employers often provide health insurance to their employees and their employees' families. There are two basic structures to health plans.

Fee-for-service plans allow patients to see doctors of their choice.

Managed care plans provide health care at a discount to members of the insured group. Health maintenance organizations (HMOs) are popular forms of managed care.

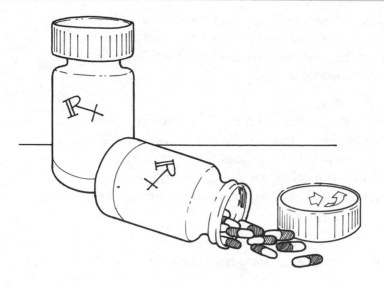

Health insurance policies often have a **co-payment** schedule. This is similar to the deductibles in car insurance and homeowner's insurance policies. For example, each visit to a doctor's office may require a co-payment from the insured of $10 to $20. Prescription drugs also require a co-payment. The high cost of prescription drugs is a serious issue for many people in the United States.

Disability insurance provides income if the policy owner is not able to work because of illness or injury. Some employers' health insurance policies include coverage for disability.

Long-term care insurance provides skilled nursing and housekeeping care to a patient in a nursing home or in his or her own house.

Life insurance is protection for the policy owner's loved ones in case of death. If the insured dies, the insurer will make a payment, called the **benefit,** to the person whom the insured has chosen—the **beneficiary.** There are two basic types of life insurance.

Term life insurance is for a fixed number of years. Term life insurance may be renewable annually, which means that each year's premium goes up. Or it may be at a level premium, which means that the price does not change. Life insurance is especially important for young families who could experience severe financial hardship if the main wage earners were to die.

Whole life insurance requires that premiums are paid for the insured's entire life.

To solve the problems in the next exercise, review:

- adding and subtracting decimals, page 230
- finding a percent of a number, page 236
- finding what percent one number is of another, page 238

EXERCISE 17

Part A

Use the following information to answer problems 1 to 3.

SITUATION

Holly has renter's insurance for her apartment. The premium is $165 a year or $16.75 a month. The policy deductible is $250.

1. How much will Holly pay in a year if she pays by the month?

2. The water from a broken pipe in Holly's upstairs neighbor's apartment destroyed Holly's computer. If the replacement cost of the computer is $1,200, how much can Holly expect from her insurance company?

3. The yearly premium for Holly's renter's insurance will rise 4% next year. What will be the yearly premium if Holly makes just one payment?

Part B

4. The premium for Mr. and Mrs. Garcia's homeowner's insurance is 0.72% of the market value of their house. The Garcias' house is worth $133,500. How much is their yearly home insurance premium?

5. Since Mr. and Mrs. Garcia in problem 4 have an automobile insurance policy with the same company that provides their house insurance, they will get a 6% discount on their home insurance premium. What is the discounted yearly home insurance premium?

6. Jim's automobile insurance premium is $847 this year. His insurance agent says that the premium is likely to rise 9% for next year. What will be next year's premium?

7. Mrs. Gillette's homeowner's insurance premium was $721 two years ago. Last year the premium was $784. By what percent did the premium rise?

8. In 1990, the average cost per hospital stay in the U.S. was $4,947. In 2000, the average cost per hospital stay was $6,649. By about what percent did the average cost of a hospital stay rise in ten years?

9. Per person health care expenditures include the cost of health insurance, out-of-pocket payments for medical services, and the cost of drugs and medical supplies. In 1990, the average expenditure per person for health care was $1,480. In 2000, the average expenditure per person was $2,066. By about what percent did the average expenditure rise in ten years?

Part C

The following table describes the coverage, limits, and premiums for Joaquin's car insurance. Use the table to answer problems 10 to 12.

		premium
Liability	$500,000/accident	$ 224
Medical	$10,000/person	$ 4
Comprehensive	$200 deductible	$ 110
Collision	$200 deductible	$ 169
Underinsured Motorists	$500,000/accident	$ 96
Personal Injury Protection	$125,000	$ 102
Towing	$50 limit	$ 6

10. What is the total yearly premium for Joaquin's car insurance?

11. If a tree falls on Joaquin's car and causes $430 in damage, how much will the insurer have to pay to make the repair?

12. The premium for liability is about what percent of the total cost of Joaquin's car insurance?

Part D

The following table tells the required co-payments for Francesca's health insurance. Use the table to answer problems 13 to 15.

Co-payment Schedule	
Office Visits	
Primary Care Provider Office Visit	$15 co-pay
Specialist Office Visit	$15 co-pay
Routine Eye Exam	$15 co-pay
Hospital Services	
Inpatient Services	$240 co-pay*
Outpatient Services with Referral	No co-pay
Emergency Services	No co-pay
Other Services	
Laboratory Services	No co-pay
Diagnostic Services incl. X-ray	$15 co-pay

*$240 co-payment is limited to two per member per calendar year per continuous hospital confinement.

13. Find the total co-payments for two visits to Francesca's primary care provider and a three-day hospital stay for minor surgery.

14. In June, Francesca went to her doctor for a routine checkup. The doctor asked her to go to another facility the following week for blood tests and an X-ray. What total co-payments did Francesca make for these examinations and procedures?

15. Francesca paid $2,976.80 for her health insurance policy. If the premium rises 8.5%, what will be her health insurance premium next year?

Part E

The following table tells the cost of monthly life insurance premiums. To find the monthly premium for a 39-year-old male, use the premiums for "Male 35." To find the monthly premium for a 52-year-old female, use the premiums for "Female 50."

Use the table to answer problems 16 to 19.

Guaranteed Level Monthly Premiums for $500,000 Death Benefit		
	20-Year	30-Year
Male 35	$ 26.55	$ 50.85
Female 35	$ 24.30	$ 39.60
Male 40	$ 40.50	$ 75.15
Female 40	$ 32.40	$ 54.00
Male 45	$ 63.00	$ 108.45
Female 45	$ 46.80	$ 73.80
Male 50	$ 97.20	N/A
Female 50	$ 72.45	N/A
Male 55	$ 151.20	N/A
Female 55	$ 105.30	N/A

16. Find the yearly premium for a 48-year-old male who wants a 20-year term benefit of $500,000.

17. What is the yearly premium for a 36-year-old female who wants a 30-year term benefit of $500,000?

18. Find the total cost over 20 years for a 20-year $500,000 benefit for a 38-year-old male.

19. How much more does it cost in a year for a 54-year-old male to pay for a 20-year $500,000 life insurance policy than it costs a 54-year-old female to pay for the same policy?

Part F

20. Sean was advised by a financial counselor that he should get a life insurance policy that would pay about six times his yearly contribution to his family's income. If he brings home $42,000 a year, how much should the benefit be?

Post-Lesson Vocabulary Reinforcement

Match the terms on the right with the definitions. Write the appropriate letters in the blanks.

_____ 1. protection against financial loss

_____ 2. a document that represents an agreement between the insured and the insurer

_____ 3. the person who offers the insurance—the insurer

_____ 4. the dollar amounts and limits that the insurer will pay

_____ 5. the insured or policyholder

_____ 6. protects a policyholder from the costs that result from damage to one's vehicle

_____ 7. the price a customer pays for insurance

_____ 8. a request that an insurance policyholder makes for payment

_____ 9. protects against the high costs of treating illnesses and injuries

_____ 10. protects a family against the loss of income that results from the policyholder's death

a. the first party

b. insurance

c. an insurance policy

d. coverage

e. the premium

f. the second party

g. automobile insurance

h. a claim

i. life insurance

j. health insurance

Choose the appropriate terms on the right to complete the sentences.

11. A _____ is the amount that the policy owner must pay on a claim.

12. _____ refers to legal obligation.

13. Insurance for the policy owner and the passengers in his car is an example of _____.

14. _____ covers the cost of damage or loss of the policy owner's car caused by something other than a collision.

15. _____ protection covers costs to the policy owner from accidents caused by a driver with inadequate insurance.

16. _____ pays for each driver's injuries regardless of who caused the accident.

17. Coverage added to one's homeowner's policy is called a

_____.

18. The federal government offers _____ for people 65 and older.

19. For people without sufficient income or resources, there is

_____.

rider

first-party coverage

deductible

no-fault insurance

comprehensive insurance

Medicare

uninsured motorist

Medicaid

liability

Language Builder

ACTIVITY A Remember that a **clause** is a kind of sentence inside another sentence. Remember that clauses begin with words like *that, who, when, until, whenever.* Underline the clauses in the following sentences. These words are called *clause markers* (CM). Double underline the clause markers in the exercise above.

1. An insurance policy is an agreement between a consumer who pays for a policy and the company that offers the insurance.

2. An insurance policy is a detailed document that describes the coverage that the insurer will pay in case of a financial loss.

3. When a policyholder makes a claim, he often has to pay a deductible.

ACTIVITY B Complete the following clauses with your own words. Then compare your work with a partner and with the class.

4. Insurance that _____ is called health insurance.

5. Someone who _____ is called the insured.

6. A co-payment requires that the insured _____.

ACTIVITY C Remember that **adverbs** are words that can modify verbs. For more information about adverbs, go to Lesson 4. Underline only the adverbs in the following sentences. Then double underline the verbs that the adverbs modify.

7. There is usually a deductible for collision insurance.

8. The value of a car depreciates quickly.

9. Term life insurance may be renewed annually.

ACTIVITY D On a separate sheet of paper, use the following adverbs in your own sentences. Then compare your work with a partner.

10. comfortably

11. annually

12. frequently

13. usually

14. carefully

LESSON 18: TAXES

Pre-Lesson Vocabulary Practice

Study the following terms and their meanings. Then find them in the lesson and read the surrounding sentences to help you better understand the meanings. Finally, take turns with a partner choosing a term from the list and giving the meaning.

source – a place where something comes from

supports – maintains, takes care of

personal income tax – tax from an individual

corporate income tax – tax from a corporation or business

imposed – charged, placed

airfare – the price of an airline ticket

segment – part, portion

assessed – determined

jurisdiction – a legally determined area

outbound – departing, leaving

stopover – a separate landing, a brief stay

return flight – air travel going back to where it started

weighs – amount of heaviness

budget – an estimate or a plan of how to spend money

Write your own sentences to show that you know the meaning of the following terms. Compare your work with a partner and then share it with the class.

1. source: _____

2. supports: _____

3. segment: _____

4. criteria: _____

5. assessed: _____

6. imposed: _____

7. budget: _____

8. flight: _____

Taxes

The income that any government uses for its operations is called **revenue.** The main source of revenue for the government is **taxes.** The revenue that comes from **property taxes** supports local governments and schools. The income from **sales tax** and **state income tax** supports state governments. The income from **personal income tax** and **corporate income tax** supports the U.S. government.

An **excise tax** is a tax imposed by a government on specific goods and services. There are excise taxes on gasoline, telephone calls, airplane trips, cigarettes, alcohol, and tires. Excise taxes support both state governments and the federal government. The mathematics of excise taxes is simple. The tax can either be a percent of the cost of something, a flat charge, or a combination of a percent and a charge.

EXAMPLE 1   In a recent year, the excise tax on airfare was 7.5% plus $3.00 for each flight segment. Aaron bought a plane ticket from San Antonio to Baltimore. The airfare was $268. The flight required a change in St. Louis. What was the excise tax on the airfare?

Solution   Find 7.5% of $268.

$0.075 \times \$268 = \$20.10$

There were two segments to Aaron's flight—one from San Antonio to St. Louis and another from St. Louis to Baltimore. There was a charge of $3.00 for each segment.

The excise tax on the airfare was

$\$20.10 + \$3.00 + \$3.00 = $ **$26.10**

EXAMPLE 2   In the state where John lives, the excise tax on gasoline is 22¢ for every gallon. In addition, the federal government charges an excise tax of 18.3¢ for every gallon of gasoline. If one gallon of gasoline costs $2.50, what percent of that price is for excise taxes?

Solution   Add to find the total excise tax per gallon.

$0.22 + $0.183 = $0.403

Excise tax is $\frac{\$0.403}{\$2.50} = 0.161\ldots$ or **about 16% of the price of one gallon.**

Land and buildings—sometimes called **real property** or **real estate**—are taxed to support local governments and schools. **Property taxes** are based on the **assessed value** of a building or a piece of land. An **assessor** is a local government official whose job is to determine the value of the properties in his jurisdiction. In some communities, the assessed value of a property is based on criteria such as location, condition, number of bathrooms, number of bedrooms, and so on.

Sometimes property tax is simply a percent of the assessed value of a property.

EXAMPLE 3   Ms. Kay's property is assessed at $104,000. What is her yearly property tax if the rate is 2.6% of the assessed value?

Solution   Find 2.6% of $104,000.

0.026 × $104,000 = **$2,704**

In other communities, property tax is based on the **market value** of a property—that is the price an owner should be able to get for the property. And sometimes the property tax rate is given as a number of dollars per $1,000 of value.

EXAMPLE 4   Mereille's house has a market value of $119,500. The property
tax rate in her community is $12.56 per $1,000 of market value.
What is her property tax for the year?

Solution   Divide $119,500 by $1,000.

$119,500 ÷ $1,000 = 119.5

Multiply 119.5 by the tax rate.

119.5 × $12.56 = **$1,500.92**

In other communities, property tax rates are measured in mills.
A **mill** is one-tenth of a cent or one-thousandth of a dollar. To change
mills to dollars, move the decimal point three places to the left. For
example, a rate of 13 mills is equal to $0.013 per dollar. The word
millage refers to a property tax rate in mills per dollar of value.

EXAMPLE 5   The assessed value of Mr. and Mrs. Martin's house is $85,000.
Their property tax rate is 19 mills for every dollar of assessed
value. What is their yearly property tax?

Solution   **Multiply the assessed value of the house by the tax rate.
19 mills is $0.019.**

85,000 × $0.019 = **$1,615.00**

Notice that in the solution to the last example, the dollar sign ($)
disappeared in the assessed value of the house. The tax rate was given in
mills per dollar of value. The assessed value, $85,000, divided by $1 is the
whole number 85,000.

A property tax rate measured in mills and a property tax rate
measured in dollars per thousand dollars of value are different ways of
saying the same thing. In both methods, the tax rate is per thousand.

To solve the problems in the next exercise, review:
- multiplying decimals, page 231
- finding a percent of a number, page 236
- finding what percent one number is of another, page 238

EXERCISE 18

Part A

Use a calculator to solve the problems.

The excise tax on airfare is 7.5% of the ticket prices plus $3.00 for each segment of the trip. Use this information to answer problems 1 to 3.

1. Melvin bought a one-way airplane ticket from Dallas to Chicago for $431. What was the excise tax for that airfare?

2. Mr. and Mrs. Gonzalez bought round-trip tickets from Charlotte, North Carolina, to Minneapolis, Minnesota. One round-trip fare was $532. What was the total price for the two of them, including excise tax?

3. Nina purchased a round-trip ticket from New York City to Los Angeles at a discount fare of $236.50. The outbound flight had a stopover in Cincinnati. The return flight had a stopover in Chicago. What was the excise tax on her airfare?

Part B

Use the following table to answer problems 4 to 8.

Gasoline Excise Tax for the Federal Government and Selected States	
	¢ per gallon
Federal	18.3
Arizona	18.0
Connecticut	25.0
Georgia	7.5
Indiana	15.0
Maryland	23.5
Montana	27.0
Rhode Island	30.0

4. Which state listed in the table has the lowest gasoline excise tax?

5. Which state listed has the highest gasoline excise tax?

6. What is the combined federal and state excise tax on a gallon of gasoline in Connecticut?

7. The state gasoline excise tax in Indiana is what fraction of the state gasoline excise tax in Rhode Island?

8. Bob paid $2.55 for a gallon of gasoline in Maryland. What percent of the price goes for excise tax?

Part C

Use the following information to answer problems 9 and 10.

SITUATION

The federal excise tax on a tire that weighs over 90 pounds is $10.50 plus $0.50 for each pound over 90.

9. The tires on Steve's truck each weigh 110 pounds. What is the federal excise tax on four new tires for his truck?

10. Miguel owns a small trucking firm. He decided to buy 50 new tires. The tires for his trucks each weigh 120 pounds. Calculate the total excise tax for the new tires.

11. Jessica's house has a market value of $127,900. Her combined yearly property taxes are 2.18% of the market value. What are the yearly property taxes?

12. Jacob and Susan's house has a market value of $149,000. They pay $1,812 a year in property taxes. Their property taxes are what percent of the market value of their house? Calculate the answer to the nearest tenth of one percent.

13. An assessor told Gordon that his new garage would add $12,500 to the assessed value of his property. If the property tax rate is 2.95% of assessed value, how much additional property tax will Gordon have to pay for the new garage?

Part D

Use the following information to answer problems 14 to 17.

SITUATION

Murry and Anette own a small shop where they make leather wallets and handbags. The tax assessor assigned a value of $78,200 to their shop. Each year they have to pay town, county, and school taxes on their shop.

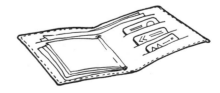

14. What are the yearly town taxes on Anette and Murry's shop if the tax rate is $4.523 per $1,000 of assessed value?

15. The rate for county taxes is $3.738 per $1,000 of assessed value. Calculate the yearly county tax for the shop.

16. The school tax rate is $7.916 per $1,000 of assessed value. What are the yearly school taxes for the shop?

17. If the school taxes rise 4.8% next year, what will be the new school taxes on Anette and Murry's shop?

Part E

SITUATION

Laura and Jack have to pay county tax and school tax on their property. The assessed value of their property is $115,300. Use this information to answer problems 18 and 19.

18. The county tax rate is 6.57 mills per $1 of assessed value. What is their yearly county tax?

19. The school tax rate for Laura and Jack's area is 14.93 mills per $1 of value. Find their yearly school taxes.

20. A rural school district has a yearly budget of $26,300,000. Property taxes fund 59% of the budget. If there are about 12,000 properties in the school district, what average amount of the year's budget comes from each property? Round your answer to the nearest hundred dollars.

Post-Lesson Vocabulary Reinforcement

Choose the appropriate terms on the right to complete the sentences.

1. The income that any government uses for its operations is called

   _____.

2. An _____ is a tax imposed by a government on specific goods and services.

3. In some communities, property tax rates are measured in

   _____.

4. _____ are based on the assessed value of a building or piece of land.

5. An _____ is a local government official whose job is to determine the value of the properties in his jurisdiction.

6. Sometimes _____ is simply a percent of the assessed value of a property.

7. The price an owner should be able to get for a property is called the

   _____.

8. A _____ is one-tenth of a cent or one-thousandth of a dollar.

9. Land and buildings are sometimes called _____.

10. The word _____ refers to a property tax rate in mills per dollar of value.

real property or real estate
property taxes
assessor
revenue
excise tax
property tax
market value
mill
millage
mills

In the blanks, write *T* if the statement is true and *F* if the statement is false. If the statement is false, rewrite the statement to make it true.

_____ 11. The main source of revenue for the government is taxes.

_____ 12. Taxes are never imposed on airfares.

_____ 13. A government can impose taxes on specific goods and services.

_____ 14. Land and buildings are taxed to support local governments and schools.

_____ 15. Property taxes are imposed on goods and services.

_____ 16. The property tax rate can be given as a number of dollars per $1,000 of value.

Now compare your answers with a partner and then correct any statements that are false. Next compare your corrections and your first answers with the class.

Language Builder

ACTIVITY A If necessary, review the **passive form** of the verb in Lessons 3 and 6. Then complete the following sentences by making the verb in parentheses passive.

1. The income that any government uses for its operations (call) _____ revenue.

2. Land and buildings (tax) _____ to support local governments and schools.

3. Ms. Kay's property (assess) _____ at $104,000.

4. Property tax can (base) _____ on the market value of a property.

5. In some communities, property tax rates (measure) _____ in mills.

6. Last year the tax rate (give) _____ in mills. (*Use the past tense here.*)

To help determine that you've used the correct form of the verb *to be* in your answers above, now <u>underline</u> the subject of the passive verb in each sentence. Then compare your answers to both parts of this exercise with a partner.

ACTIVITY B The following sentences from this lesson have been rewritten in the **present tense.** Change the verbs in **boldfaced type** to the simple past tense and write the new sentences on a separate sheet of paper. Notice that all the verbs are irregular.

7. Mr. and Mrs. Gonzalez **buy** round-trip tickets from Charlotte, North Carolina, to Minneapolis, Minnesota.

8. One round-trip fare **is** $532.

9. The return flight **has** a stopover in Chicago.

10. Bob **pays** $2.55 for a gallon of gasoline in Maryland.

ACTIVITY C The following sentences are in the **past tense.** Change the verbs in **boldfaced type** to the **simple present tense.** Then compare your answers with a partner.

11. The dollar sign ($) **disappeared** in that stated assessed value of the Martin's house.

12. The property tax rate **was** given as a number of dollars per $1,000 of value.

13. This lesson **included** examples of excise taxes and property taxes.

14. There **were** two segments to Aaron's flight.

15. How much **did** the owner get for the property?

16. Nina **paid** a lot for travel.

LESSON 19: INCOME TAX

Pre-Lesson Vocabulary Practice

Study the terms on the right and their meanings. Then find them in the lesson and read the surrounding sentences to see if you better understand the meanings. Finally, take turns with a partner choosing a term from the list and then giving the meaning.

Write your own sentences to show that you know the meanings of the following terms. Compare your work with a partner and then share it with the class.

1. extension: _____

   _____

2. to file: _____

   _____

3. spouse: _____

   _____

4. dependents: _____

   _____

5. itemized: _____

   _____

6. allowable: _____

   _____

7. status: _____

   _____

8. self-employed: _____

   _____

9. a refund: _____

   _____

extension – additional time

to file – to send in, submit

spouse – a partner in a marriage—a husband or a wife

alimony – money paid to a dependent ex-spouse

dependents – children

itemized – listed separately

allowable – something that is allowed, permitted

status – situation, condition

filing jointly – both husband and wife filing (taxes) as a unit

self-employment – when one works for oneself

credits – points for the taxpayer so he or she has to pay less

a refund – a return of one's money

penalties – amount of money charged for overdue payments

couple – husband and wife

Income Tax

Almost half of the revenue of the U.S. government comes from individual **income taxes.** The **Internal Revenue Service** (IRS) is the government agency responsible for collecting income tax. Every April 15 is the last day—unless the taxpayer applies for an extension—for a taxpayer to **file** a **tax return** for the previous calendar year. The return is an official document (1040EZ, 1040A, or 1040) that lists income, exemptions, deductions, and the tax that is due.

Filling out tax forms isn't fun, but it is an unavoidable responsibility. The mathematics of calculating taxes is fairly simple: add some numbers, subtract some numbers, look up a number in a table, or find a percent of a number. This lesson will illustrate only the most basic steps in calculating income taxes.

- First, calculate **gross income.** This is the sum of all the money a taxpayer made in a year. Gross income includes wages, salary, interest income, stock dividends, tips, alimony income, rent income, and any business income.

- Next, calculate **adjusted gross income.** From the gross income, subtract contributions to any retirement accounts, alimony payments, and interest payments on student loans.

- Then find the **taxable income** for the year. From the adjusted gross income, subtract allowable **exemptions** and **deductions.** An exemption is an amount of income that is free from taxation. The dollar amount for an exemption changes each year. A taxpayer gets an exemption for himself, for his spouse, and for any **dependents** he supports.

- Deductions may be **itemized** or **standard.** Each allowable itemized deduction must be described and listed on a separate form (Schedule A). Itemized deductions include medical and dental expenses, state and local taxes, real estate taxes, mortgage interest, charitable gifts, and job-related expenses. A standard deduction is a number determined by the IRS each year. The taxpayer may subtract the total of the itemized deductions or the standard deduction—whichever is larger—from the taxable income.

- Look up the **tax due** in a table or calculate the tax from a tax rate schedule. The tax due depends on the amount of the taxable income and on the filing **status** of the taxpayer—single, married filing jointly, married filing separately, or head of household. Then add any other taxes such as self-employment taxes and subtract any credits such as those for children or the elderly.

- Finally, compare this last number with the tax that was already paid during the year. For most employees, income tax is withheld from each paycheck. (You looked at withholding tables in the first lesson in this book.) Self-employed people usually pay **estimated taxes** during the year. The taxpayer either owes more money or can expect a **refund.**

The examples and exercises in this lesson use tables and schedules for the year 2003. Remember that the tables and percentages change from year to year.

EXAMPLE 1 Bill's gross income for the year was $31,100. He made a contribution of $2,000 to his retirement account. Find Bill's adjusted gross income for the year.

Solution **Subtract his IRA contribution (an adjustment) from his gross income.**

$31,100 – $2,000 = **$29,100**

EXAMPLE 2 Bill took a standard deduction of $4,750 and the allowable exemption of $3,050. Find his taxable income.

Solution **Subtract the standard deduction and the exemption from Bill's adjusted gross income.**

$29,100 – $4,750 – $3,050 = **$21,300**

There are two ways to find the amount of tax that Bill owes. One way is to use a tax rate schedule.

Partial 2003 Tax Rate Schedules

	If TAXABLE INCOME		The TAX is			
			THEN			
	is Over	But Not Over	This Amount	Plus This %		Of the Excess Over
SCHEDULE X—						
Single	$0	$7,000	$0.00	10%		$0.00
	$7,000	$28,400	$700.00	15%		$7,000
	$28,400	$68,800	$3,910.00	25%		$28,400
	$68,800	$143,500	$14,010.00	28%		$68,800
SCHEDULE Y-1—						
Married Filing Jointly or Qualifying Widow(er)	$0	$14,000	$0.00	10%		$0.00
	$14,000	$56,800	$1,400.00	15%		$14,000
	$56,800	$114,650	$7,820.00	25%		$56,800
	$114,650	$174,700	$22,282.50	28%		$114,650
SCHEDULE Y-2—						
Married Filing Seperately	$0	$7,000	$0.00	10%		$0.00
	$7,000	$28,400	$700.00	15%		$7,000
	$28,400	$57,325	$3,910.00	25%		$28,400
	$57,325	$87,350	$11,141.25	28%		$57,325
SCHEDULE Z—						
Head of Household	$0	$10,000	$0.00	10%		$0.00
	$10,000	$38,050	$1,000.00	15%		$10,000
	$38,050	$98,250	$5,207.50	25%		$38,050
	$98,250	$159,100	$20,257.50	28%		$98,250

EXAMPLE 3 Bill's filing status is single. How much income tax does he owe?

Solution   Bill's taxable income is $21,300. According to Schedule X for single taxpayers, his taxable income is over $7,000 but not over $28,400. He owes $700.00 plus 15% of the excess over $7,000.

The excess over $7,000 is $21,300 − $7,000 = $14,300.

15% of $14,300 = 0.15 × $14,300 = $2,145

Bill's tax is $700 + $2,145 = **$2,845.**

The other way to find tax is to use a tax table. In the table below, "line 6" refers to the taxable income line on form 1040EZ for 2003.

EXAMPLE 4 Use the tax table to find Bill's tax if his taxable income is $21,300.

Solution   In the table, find the line that is "at least 21,300 but less than 21,350." Look for the tax under the label "single." Bill's tax is **$2,849.**

Notice that the two methods do not give exactly the same results. The tax from the rate schedule is $4 less than the tax from the table.

If Form 1040EZ, line 6, is—		And you are—	
At least	But less than	Single	Married filing jointly
		Your tax is—	
21,000			
21,000	21,050	2,804	2,454
21,050	21,100	2,811	2,461
21,100	21,150	2,819	2,469
21,150	21,200	2,826	2,476
21,200	21,250	2,834	2,484
21,250	21,300	2,841	2,491
21,300	21,350	2,849	2,499
21,350	21,400	2,856	2,506
21,400	21,450	2,864	2,514
21,450	21,500	2,871	2,521
21,500	21,550	2,879	2,529
21,550	21,600	2,886	2,536
21,600	21,650	2,894	2,544
21,650	21,700	2,901	2,551
21,700	21,750	2,909	2,559
21,750	21,800	2,916	2,566
21,800	21,850	2,924	2,574
21,850	21,900	2,931	2,581
21,900	21,950	2,939	2,589
21,950	22,000	2,946	2,596

If Bill has no credits or penalties, there is only one more step to determine the **tax that is due** to the IRS or the amount of his **refund.** Compare the amount of tax that Bill owes with the amount that was withheld from his pay for the year or with the amount of estimated tax that he paid during the year.

EXAMPLE 5   According to his year-end wage statement (form W-2), Bill's employer withheld a total of $3,400 from his paychecks during the year. Find the amount that Bill has to send to the IRS or the amount of his refund.

Solution   Since the tax that was withheld from his pay is greater than the amount that is due, Bill will get a refund.

$3,400 − $2,849 = **$551**

The five examples illustrate a very simple tax case. Bill's tax situation can be summarized on the simplest of the IRS forms—**1040EZ** (EZ stands for easy). Taxpayers with more complicated situations should use Form **1040A** or Form **1040.** Taxpayers who realize that they sent in a form with wrong information can make corrections on Form **1040X.**

Many states collect personal income tax as well. A few states, like Florida and Texas, have no state income tax. Other states, like Colorado and Rhode Island, simply charge a percent of the federal income tax.

There are many places to get help in filling out tax forms. The IRS publishes detailed booklets that explain the various forms. Information about state taxes is on page 208 in this book. In most communities there are professionals who can help taxpayers complete their tax forms. Increasingly, taxpayers use computer programs to prepare their taxes.

To solve the problems in the next exercise, review:
- finding a percent of a number, page 236

EXERCISE 19

Part A

The table below lists the tax brackets for 2003. Use the table to answer problems 1 to 5.

1. What is the highest rate listed on the table?

2. What is the tax rate for a single taxpayer with a taxable income of $45,000?

3. What is the tax rate for a couple filing a joint return if their taxable income is $29,000?

4. What is the lowest taxable income for a joint return that is taxed at 33%?

5. What is the highest taxable income for a single taxpayer at the 15% rate?

Tax Brackets—2003 Taxable Income

Joint return	Single taxpayer	Rate
$0–$14,000	$0–$7,000	10.0%
14,000–56,800	7,000–28,400	15.0
56,800–114,650	28,400–68,800	25.0
114,650–174,700	68,800–143,500	28.0
174,700–311,950	143,500–311,950	33.0
311,950 and up	311,950 and up	35.0

Part B

For problems 6 to 10, use the 2003 Partial Tax Rate Schedules on page 200. Identify the schedule (X, Y-1, Y-2, or Z) that you need, and then solve each problem.

6. Joe and Ann Camden are married and file their tax return jointly. Their taxable income in 2003 was $63,490. What was their federal income tax for the year?

7. Mavis files her income tax as a head of household. Her 2003 taxable income was $29,215. According to the tax rate schedule, what amount of tax did she owe?

8. Janet Johnson is married, but she and her husband file their income taxes separately. Her taxable income was $43,760. How much tax did she owe?

9. Martin made $72,500 in taxable income in 2003. If he is single, how much tax did he owe?

10. Mr. and Mrs. Cozzi file their income taxes jointly. If their taxable income for the year was $62,850, how much tax did they owe?

Part C

In the tax table below, "line 40" refers to the taxable income line on Form 1040.
Use the table to answer problems 11 to 20.

11. Juanita is single. Her taxable income for 2003 was $32,771. How much federal income tax did she owe for the year?

12. Juanita, in the last problem, paid estimated taxes of $4,500 in 2003. Calculate the amount that she still had to pay the IRS or the amount of her refund.

13. Mr. Armstrong is a head of household. He made $33,208 in 2003. According to the tax table, how much tax did he have to pay?

14. Mr. Armstrong, in the last problem, had $5,200 withheld from his pay in 2003 for federal income taxes. Find the additional amount that he had to pay or the amount of his refund.

15. Sue and Tom Madison are married and filing their taxes jointly. If their taxable income was $32,664, how much tax did they owe?

2003 Tax Table—

If line 40 (taxable income) is—		And you are—			
At least	But less than	Single	Married filing jointly	Married filing separately	Head of a household
			Your tax is—		
32,000					
32,000	32,050	4,816	4,104	4,816	4,304
32,050	32,100	4,829	4,111	4,829	4,311
32,100	32,150	4,841	4,119	4,841	4,319
32,150	32,200	4,854	4,126	4,854	4,326
32,200	32,250	4,866	4,134	4,866	4,334
32,250	32,300	4,879	4,141	4,879	4,341
32,300	32,350	4,891	4,149	4,891	4,349
32,350	32,400	4,904	4,156	4,904	4,356
32,400	32,450	4,916	4,164	4,916	4,364
32,450	32,500	4,929	4,171	4,929	4,371
32,500	32,550	4,941	4,179	4,941	4,379
32,550	32,600	4,954	4,186	4,954	4,386
32,600	32,650	4,966	4,194	4,966	4,394
32,650	32,700	4,979	4,201	4,979	4,401
32,700	32,750	4,991	4,209	4,991	4,409
32,750	32,800	5,004	4,216	5,004	4,416
32,800	32,850	5,016	4,224	5,016	4,424
32,850	32,900	5,029	4,231	5,029	4,431
32,900	32,950	5,041	4,239	5,041	4,439
32,950	33,000	5,054	4,246	5,054	4,446
33,000					
33,000	33,050	5,066	4,254	5,066	4,454
33,050	33,100	5,079	4,261	5,079	4,461
33,100	33,150	5,091	4,269	5,091	4,469
33,150	33,200	5,104	4,276	5,104	4,476
33,200	33,250	5,116	4,284	5,116	4,484
33,250	33,300	5,129	4,291	5,129	4,491
33,300	33,350	5,141	4,299	5,141	4,499
33,350	33,400	5,154	4,306	5,154	4,506
33,400	33,450	5,166	4,314	5,166	4,514
33,450	33,500	5,179	4,321	5,179	4,521
33,500	33,550	5,191	4,329	5,191	4,529
33,550	33,600	5,204	4,336	5,204	4,536
33,600	33,650	5,216	4,344	5,216	4,544
33,650	33,700	5,229	4,351	5,229	4,551
33,700	33,750	5,241	4,359	5,241	4,559
33,750	33,800	5,254	4,366	5,254	4,566
33,800	33,850	5,266	4,374	5,266	4,574
33,850	33,900	5,279	4,381	5,279	4,581
33,900	33,950	5,291	4,389	5,291	4,589
33,950	34,000	5,304	4,396	5,304	4,596

16. The Madisons, from problem 15, paid estimated taxes of $4,150 in 2003. Find the amount that they still owed the IRS or the amount of their refund.

17. Melinda is single. Her taxable income for the year was $32,989. How much tax did she owe?

18. Melinda, in the last problem, had a total of $4,980 withheld from her paychecks for federal income tax. Calculate the additional amount that she owed or the amount of her refund.

19. Sam is married, but he and his wife file their income taxes separately. His taxable income for the year was $33,347. How much tax did he owe?

20. Sam, in the last problem, had $5,700 withheld from his pay for the year. Find the additional amount that he owed or the amount of his refund.

Part D

21. Use the following information about James Gordon, a bus driver, to fill out Form 1040EZ on the next page.

> **James Q. Gordon**
> **625 Maple Terrace, Apt. 3F**
> **Fairmont, NY 12565-0625**
>
> **Social security number 222-99-3333**

Use the information above to fill out the "Label" section. Decide for yourself whether James wishes to contribute to the Presidential Election Campaign.

For the "Income" section, James' salary was $32,410 for the year, and his taxable interest was $112. He had no unemployment compensation or dividends. His taxpayer status is single. Follow the instructions on lines 4 and 6.

For the "Payments and tax" section, James' employer reported on a W-2 form that he withheld a total of $3,640 from James' paychecks for the year. The earned income credit does not apply to James.

Use the tax table at the right to find James' tax.

Decide whether James will get a refund or have to pay more tax.

If Form 1040EZ, line 6, is—		And you are—	
At least	But less than	Single	Married filing jointly
		Your tax is—	
24,000			
24,000	24,050	3,254	2,904
24,050	24,100	3,261	2,911
24,100	24,150	3,269	2,919
24,150	24,200	3,276	2,926
24,200	24,250	3,284	2,934
24,250	24,300	3,291	2,941
24,300	24,350	3,299	2,949
24,350	24,400	3,306	2,956
24,400	24,450	3,314	2,964
24,450	24,500	3,321	2,971
24,500	24,550	3,329	2,979
24,550	24,600	3,336	2,986
24,600	24,650	3,344	2,994
24,650	24,700	3,351	3,001
24,700	24,750	3,359	3,009
24,750	24,800	3,366	3,016
24,800	24,850	3,374	3,024
24,850	24,900	3,381	3,031
24,900	24,950	3,389	3,039
24,950	25,000	3,396	3,046

Form
1040EZ

Department of the Treasury—Internal Revenue Service

Income Tax Return for Single and Joint Filers With No Dependents (99) **2003**

OMB No. 1545-0675

Label

(See page 12.)

Use the IRS label. Otherwise, please print or type.

L A B E L   H E R E

Your first name and initial	Last name	Your social security number
If a joint return, spouse's first name and initial	Last name	Spouse's social security number
Home address (number and street). If you have a P.O. box, see page 12.		Apt. no.
City, town or post office, state, and ZIP code. If you have a foreign address, see page 12.		

▲ **Important!** ▲

You **must** enter your SSN(s) above.

Presidential Election Campaign (page 12) ▶

Note. Checking "Yes" will not change your tax or reduce your refund.

Do you, or your spouse if a joint return, want $3 to go to this fund? . . . . . . ▶

	You		Spouse	
	☐ Yes	☐ No	☐ Yes	☐ No

Income

Attach Form(s) W-2 here. Enclose, but do not attach, any payment.

1 Wages, salaries, and tips. This should be shown in box 1 of your Form(s) W-2. Attach your Form(s) W-2.   **1**

2 Taxable interest. If the total is over $1,500, you cannot use Form 1040EZ.   **2**

3 Unemployment compensation and Alaska Permanent Fund dividends (see page 14).   **3**

4 Add lines 1, 2, and 3. This is your **adjusted gross income.**   **4**

Note. You **must** check Yes or No. }

5 Can your parents (or someone else) claim you on their return?

Yes. Enter amount from ☐ worksheet on back.

No. If **single,** enter $7,800. ☐ If **married filing jointly,** enter $15,600. See back for explanation.   **5**

6 Subtract line 5 from line 4. If line 5 is larger than line 4, enter -0-. This is your **taxable income.** ▶   **6**

Payments and tax

7 Federal income tax withheld from box 2 of your Form(s) W-2.   **7**

8 **Earned income credit (EIC).**   **8**

9 Add lines 7 and 8. These are your **total payments.** ▶   **9**

10 **Tax.** Use the amount on **line 6 above** to find your tax in the tax table on pages 24–28 of the booklet. Then, enter the tax from the table on this line.   **10**

Refund

Have it directly deposited! See page 19 and fill in 11b, 11c, and 11d.

11a If line 9 is larger than line 10, subtract line 10 from line 9. This is your **refund.** ▶   **11a**

▶ **b** Routing number [ ][ ][ ][ ][ ][ ][ ][ ][ ]   ▶ **c** Type: ☐ Checking ☐ Savings

▶ **d** Account number [ ][ ][ ][ ][ ][ ][ ][ ][ ][ ][ ][ ][ ][ ][ ][ ][ ]

Amount you owe

12 If line 10 is larger than line 9, subtract line 9 from line 10. This is the **amount you owe.** For details on how to pay, see page 20. ▶   **12**

Third party designee

Do you want to allow another person to discuss this return with the IRS (see page 20)? ☐ **Yes.** Complete the following. ☐ **No**

Designee's name ▶	Phone no. ▶ ( )	Personal identification number (PIN) ▶ [ ][ ][ ][ ][ ]

Sign here

Joint return? See page 11.

Keep a copy for your records.

Under penalties of perjury, I declare that I have examined this return, and to the best of my knowledge and belief, it is true, correct, and accurately lists all amounts and sources of income I received during the tax year. Declaration of preparer (other than the taxpayer) is based on all information of which the preparer has any knowledge.

Your signature	Date	Your occupation	Daytime phone number ( )
Spouse's signature. If a joint return, **both** must sign.	Date	Spouse's occupation	

Paid preparer's use only

Preparer's signature ▶	Date	Check if self-employed ☐	Preparer's SSN or PTIN
Firm's name (or yours if self-employed), address, and ZIP code		EIN	
		Phone no. ( )	

For Disclosure, Privacy Act, and Paperwork Reduction Act Notice, see page 23.   Cat. No. 11329W   Form **1040EZ** (2003)

Online Tax Resources

State	Website
Alabama	www.ador.state.al.us/incometax/ITindex2.html
Alaska	www.revenue.state.ak.us
Arizona	www.revenue.state.az.us
Arkansas	www.ark.org/dfa/taxes/index.html
California	www.taxes.ca.gov
Colorado	www.colorado.gov/CS/Satellite/Revenue/REVX/1176842266433
Connecticut	www.ct.gov/drs/cwp/view.asp?a=1509&q=443198
Deleware	revenue.delaware.gov/
District of Columbia	otr.cfo.dc.gov/otr/site/default.asp
Florida	dor.myflorida.com/dor/taxes/
Georgia	www.stateofgeorgia.com
Hawaii	www.state.hi.us/tax/tax.html
Idaho	tax.idaho.gov/index.cfm
Illinois	www.revenue.state.il.us
Indiana	www.in.gov/dor
Iowa	www.iowa.gov/tax/forms/loadform.html
Kansas	www.ksrevenue.org/forms.htm
Kentucky	revenue.ky.gov/forms/
Louisiana	www.rev.state.la.us
Maine	www.state.me.us/revenue/forms
Maryland	www.comp.state.md.us
Massachusetts	www.dor.state.ma.us/forms/formsIndex/taxformsPERSONAL.htm
Michigan	www.michigan.gov/treasury
Minnesota	taxes.state.mn.us/pages/current_forms.aspx
Mississippi	www.dor.ms.gov/getincometaxforms.htm
Missouri	dor.mo.gov/personal/
Montana	revenue.mt.gov/formsandresources/default.mcpx
Nebraska	www.revenue.ne.gov/tax/forms.html
Nevada	tax.state.nv.us
New Hampshire	www.state.nh.us/revenue/forms/index.htm
New Jersey	www.state.nj.us/treasury/taxation/index.shtml
New Mexico	www.state.nm.us/tax/trd_form.htm
New York	www.tax.state.ny.us
North Carolina	www.dor.state.nc.us/forms
North Dakota	www.nd.gov/tax/indincome/forms/
Ohio	tax.ohio.gov/forms/index.stm
Oklahoma	www.oktax.state.ok.us
Oregon	www.oregon.gov/DOR/forms.shtml
Pennsylvania	www.revenue.state.pa.us
Rhode Island	www.tax.state.ri.us/taxforms/
South Carolina	www.sctax.org
South Dakota	www.sd.gov/servicedirect/Results.aspx?Cat=TAXES
Tennessee	www.state.tn.us/revenue/forms/index.htm
Texas	www.cpa.state.tx.us/taxinfo/taxforms/00-forms.html
Utah	tax.utah.gov/
Vermont	tax.vermont.gov/forms.shtml
Virginia	www.tax.virginia.gov/
Washington	dor.wa.gov/content/getaformorpublication/
West Virginia	www.state.wv.us/taxrev/personal.html
Wisconsin	www.dor.state.wi.us
Wyoming	revenue.state.wy.us/PortalVBVS/DesktopDefault.aspx?tabindex=2&tabid=9

Post-Lesson Vocabulary Reinforcement

Match the definitions with the terms on the right. Write the appropriate letters in the blanks.

_____ 1. government agency responsible for collecting income tax

_____ 2. official document that lists income, exemptions, deductions, and tax due

_____ 3. sum of all the money a taxpayer made in a year

_____ 4. gross income less contributions to any retirement accounts, alimony payments, and interest payments on student loans

_____ 5. number determined by the IRS each year

_____ 6. single, married filing jointly, married filing separately, or head of household

_____ 7. year-end wage statement from an employer

_____ 8. tax form that's supposed to be "easy" to fill out

a. gross income

b. the IRS

c. a tax return

d. 1040EZ

e. adjusted gross income

f. form W-2

g. filing status

h. standard deduction

Choose the appropriate terms from the right to complete the sentences.

9. _____ stands for Internal Revenue Service.

10. If a taxpayer needs additional time to file a return, he or she applies for an _____.

11. The 1040, 1040A, and the 1040EZ are _____.

12. Tax deductions may be _____ or standard.

13. Itemized deductions include _____.

14. The _____ depends on the amount of taxable income and one's filing status.

15. After filing, the taxpayer can expect to receive a _____ or to owe more money.

16. You can use _____ to make corrections to your original filing.

itemized

Form 1040X

charitable gifts

refund

extension

tax due

tax forms

IRS

Use the information from this lesson to answer the following information questions.

17. How do you calculate adjusted gross income?

18. Who withholds taxes from employees' paychecks?

19. When is the last date for a taxpayer to file a tax return?

20. What is an example of a tax deduction?

Language Builder

ACTIVITY A **Conjunctions** are words that connect things. Some examples are the words *and, but,* and *or.* Here are examples:

James' salary was $32,410 for the year, **and** his taxable interest was $112.

Sam is married, **but** he and his wife file their income taxes separately.

Look up the tax due in a table **or** calculate the tax from a tax rate schedule.

Note how these conjunctions work: *and* is used to join two similar things; *but* is used to join two different things; *or* is used to include an alternative—one thing instead of another.

Fill the blanks in the following sentences with *and, but,* or *or.*

1. Add any other taxes such as self-employment taxes _____ subtract any credits such as those for children or the elderly.

2. Either Bill owes more money, _____ he can expect a refund.

3. Filling out tax forms isn't fun, _____ it is an unavoidable responsibility.

4. Identify the schedule (X, Y-1, Y-2, or Z) that you need, _____ then solve each problem.

5. Joe can subtract itemized deductions, _____ he can use the standard deduction.

6. A taxpayer can apply for an extension, _____ eventually he has to pay his taxes.

ACTIVITY B Remember from Lesson 8 that a **clause,** like a main sentence, has a S = Subject + V = Verb:

> *The term means **that the owner of a stock has** an "equitable claim" on the company.*
>
> S   V   **CM**   **S**   **V**

(In the exercise that follows you will notice that not all clauses have clause markers.)

Put the words for each clause (in parentheses) in the correct order in the following sentences:

7. The tax return is an official document (lists   the   income   that   tax   and   the   due   is   that).

8. Gross income is the sum of all the money (made   a year   in   a taxpayer).

9. Finally, compare this last number with the tax (was   the year   already paid   during   that).

10. (Martin   single   If   is), how much tax does he owe?

11. In many communities there are trained professionals (to complete   can help   who   tax payers   tax forms   their).

UNIT 4
BUDGETING

This final lesson serves as a review of much of the material in the book. You will analyze some average budgets of American families, and you will get a chance to make a budget for yourself.

Writing a Budget

Most people know what their income is from each paycheck. Fewer people know exactly how they spend their income. A **budget** is a detailed summary of estimated expenses over a period of time. Governments are required by law to prepare yearly budgets. Most consumers do not bother making a budget for themselves, but it is a good idea.

A **long-term** budget helps plan for a major expense in the future such as a down payment on a house, a vacation, a new kitchen appliance, or college tuition. Every long-term budget includes some guesswork. For example, it is easy to find the cost of tuition for the current year, but it is impossible to know what tuition will be in four or five years. A long-term budget is a useful planning tool even if it is an estimate.

EXAMPLE 1   Mr. and Mrs. Acevedo hope to buy a house in four years. They want to be able to make a down payment of $15,000 when they buy the house. If they start saving now, how much do they need to save each year to accumulate the down payment?

Solution   **Divide the down payment by 4.**

$15,000 ÷ 4 = **$3,750**

EXAMPLE 2   The Acevedos, in the last example, want to put money aside each month for their down payment. How much should they save each month if they give themselves four years to accumulate the total down payment?

Solution   **Divide the amount they need to save each year by 12.**

$3,750 ÷ 12 = **$312.50**

A **short-term** budget is a financial plan for immediate and repeating expenses. A budget that includes only housing costs, food costs, and "other" is less useful than a more detailed list.

Below is a table of the average expenditures of households in the Midwest for a recent year.

Average Household Expenditures	
Housing	31.7%
Transportation	18.7
Food	13.6
Clothing	4.5
Healthcare	5.8
Insurance & pensions	9.5
Entertainment	5.6
Other	10.6

Remember that the percentages in the table of expenditures are *averages.* Individual cases vary widely. A family that owns a home and pays no monthly mortgage will probably spend a smaller percent of income on housing than a family that rents or that has just purchased a home. A worker who can walk to her job will probably spend much less on transportation than another worker who has to drive twenty miles to get to work. Car insurance is a modest expense in some parts of the country and an expensive item in others.

EXAMPLE 3 Ronald pays $524 a month for rent that includes heat. His monthly electricity bill is about $38, and his monthly phone bill is about $62. He has no renter's insurance or any other expenses for maintaining his home. If his yearly net pay (take-home pay) is $26,780, what percent of his income goes for housing?

Solution To find Ronald's total housing expenses for one month, add his monthly rent, electricity bill, and telephone bill:

$524 + $38 + $62 = $624

Next, find Ronald's monthly net pay:

$26,780 ÷ 12 = $2,231.67

Housing costs are $624 ÷ $2,231.67 = 0.2796. . .
or **about 28.0%** of his income.

To solve the last problem, you could also multiply each of his monthly housing expenses by 12. Then compare the total with his yearly net pay. The answer will be the same. Notice that Ronald's housing expenses are lower than the average percent in the table on page 213.

EXAMPLE 4   Ronald kept close track of his grocery bills for three months. He calculated that he spent an average of about $214 a month on groceries. After going through his credit card bills for three months, he found that he spent an average of $362 a month on meals in restaurants. If these were his only expenses for food, what percent of his income goes for food?

Solution   To find Ronald's average total food costs for one month, add his average monthly grocery bill and the average monthly restaurant bills:

$214 + $362 = $576

Food costs are $576 ÷ $2,231.67 = **25.8%** of his income.

Notice that Ronald's food costs are well above the average in the table on page 213. This suggests that Ronald should consider cutting back on the number of meals he eats out. Remember also that if he does not pay off his total credit balance each month, these costs will be even higher.

To solve the problems in the next exercise, review:

- adding decimals, page 230

- finding what percent one number is of another, page 238

EXERCISE 20

Use a calculator to solve the following problems.

Part A

Use the price list below to solve problems 1 to 4.

```
26-cubic-foot, side-by-side refrigerator ........$  750
48-inch, high-definition, flat screen TV .......$1,400
Stereo system with 600-watt receiver.........$1,100
Self-cleaning, smooth top electric range......$  560
```

1. If Jim saves $40 a month, how long will it take him to buy the stereo system with the 600-watt receiver?

2. In a year and a half, Silvia wants to have the cash to replace her old refrigerator with the side-by-side model listed above. How much does she have to save each month?

3. Mrs. Melville wants to replace her stove with the self-cleaning range listed above. If she puts aside $35 a month, how long will she need to save to have enough cash to buy the range?

4. For his thirtieth birthday, Karim wants to treat himself to the flat screen TV listed above. Karim just turned 28. How much does he have to save each month to have enough cash to buy the TV?

Part B

Below is a table that compares the costs of vacations in different states for a recent year. Use the table to answer problems 5 to 9.

State	Meals $	Lodging $	Total $
California	132	136	268
District of Columbia	136	226	362
Florida	112	140	252
Hawaii	144	273	417
Nevada	151	100	251
New York	135	144	279
North Dakota	89	78	167
Vermont	111	121	232

Average Daily Vacation Costs
Meals and Lodging in Selected States for Two Adults and Two Children

5. Which state on the list has the most expensive average cost for meals?

6. Which state on the list has the least expensive total for meals and lodging?

7. Phil and Sue Ellen want to take their two kids to Washington, D.C. for five nights during the kids' spring break in April. They figure that they will spend about $80 in gasoline and about $30 a day in incidental expenses. If they start saving in September, how much should they put aside each month to have enough cash by the end of March for the vacation?

8. The average cost of meals and lodging for a family of four on a two-week vacation is how much more in Hawaii than in New York?

9. Paulo and Marie plan to spend seven nights at a hotel in Florida with their two children. Paulo used the table to estimate the cost of meals and lodging. In addition, he expects to spend the following:

$200 per person for
   round-trip airfare
$20 including tip for taxi ride
   to hotel in Washington
$20 for taxi ride back to airport
   in Washington
$40 tips for hotel staff
$10 a day for long-term parking
   lot at airport
$30 a day for incidental expenses

If Paulo and Marie save for twelve months, how much should they put aside each month to have enough cash for their trip?

Part C

10. Ruth's daughter Olivia plans to go to college in five years. The tuition at the largest state university in their state is now $4,350 a year. Ruth wants to save enough to pay for Olivia's tuition. Olivia plans to save money from summer jobs and part-time work to pay for her room and board. Ruth assumes that the current tuition will rise, but she would like to save enough to have four years' tuition when Olivia starts college. If she saves for five years, how much should Ruth put aside each month to cover four years of tuition at the current rate?

Part D

Use the following information to answer problems 11 to 14.

SITUATION

> Mark and Karen want to make a budget for their expenses. They
> pay $639 each month for their mortgage. They pay a total of $816 in
> a year for heating oil. Their average monthly electricity bill is $49,
> and their average monthly telephone bill is $64. Their homeowner's
> insurance costs $689 a year, and their yearly real estate taxes are
> $1,628. Their combined take-home pay is $36,780.

11. What are Mark and Karen's average housing expenses for one month?

12. What is Mark and Karen's average monthly income?

13. What percent of their income do Mark and Karen pay for housing?

14. Mark and Karen spend $431 a month on their car loan, $847 each year on car
    insurance, a total of $528 for gasoline and oil, and about $80 for fees and highway
    tolls. What percent of their income do they spend on transportation?

Part E

Below are the expenses for Rhonda and Malcolm for one year. Use these expenses to answer problems 15 to 19.

Housing	$9,605
Transportation	$4,940
Food	$6,365
Clothes	$2,483
Healthcare	$1,691
Entertainment	$1,988
Insurance & pensions	$2,073
Other	$3,316

15. Find the total of Rhonda and Malcolm's expenses for the year.

16. If a total of $6,230 is deducted from their gross income for the year, what is their gross income?

17. To the nearest tenth, what percent of their expenses was spent in each category?

18. All together, housing, transportation, and food are what percent of Rhonda and Malcolm's budget?

19. In which categories did Rhonda and Malcolm spend more than the average percentages in the budget on page 213? In which categories did they spend less?

A Final Word:

A budget is most useful when it is personal. The list on the following page has more topics than you have seen so far in the budgets in this lesson. Use the list to create a budget for yourself and your family. Add any categories that apply to you but that you cannot find on the list. After each general category there is a list of the lessons in this book where you can find suggestions for expenses that you forgot or where you can look up vocabulary and review mathematical procedures.

Every budget needs a category named "other" or "miscellaneous." This is the place where you can include your worst habits as a consumer—such as the credit card interest that you did not pay off immediately. You can also include your best habits, such as the donations you made to charities. Don't forget to include long-range plans. If there is not enough money left for tuition or a vacation or a new kitchen, study your budget carefully. There will be some categories where you can cut back. Put the money you save in a fund for your long-term goals.

Budget Topics

HOUSING
(Lessons 1, 13, 14, 15, 17)

Mortgage or rent

Gas

Water

Electricity

Telephone

Trash pickup

Property tax

Homeowner's or renter's
  insurance

Repairs & maintenance

TRANSPORTATION
(Lessons 6, 10, 16, 17)

Car payments

Gasoline & oil

Maintenance & repairs

Fees & tolls

Car insurance

Public transportation

FOOD
(Lessons 4, 5, 12)

Groceries

Eating out

CLOTHING
(Lessons 2, 3, 12)

Personal clothing

Work clothing

Children's clothing

HEALTHCARE
(Lesson 17)

Insurance

Medicine

Dentist

Eye doctor & glasses

ENTERTAINMENT
(Lessons 5, 12)

Cable TV

Computer & Internet access

Newspapers, magazines, & books

Movies, concerts, & sports

OTHER
(Lessons 6, 8, 9, 10, 11, 12)

Credit card interest & fees

Child care

Toiletries & household items

Gifts & donations

Pets

Savings

Long-term goals

Business expenses

UNIT 5
MATH SKILLS
REVIEW

Math Skills Review

In most of the exercises in this book, there are instructions to use a calculator to solve problems. Many of the examples in this skills review illustrate solutions with a calculator.

Place Value

Numbers are written with ten **digits:** 0, 1, 2, 3, 4, 5, 6, 7, 8, and 9. The value of a digit depends on its place in a number. The 7 in 3,275 has a value of 70 because it is in the tens place. The 7 in 37,189 has a value of 7,000. The 7 in 4.7 has a value of $\frac{7}{10}$. The value of the 7 in each number depends on its **place.**

A **decimal point** separates whole numbers from decimals. The chart below lists the names of seven whole number places and six decimal places.

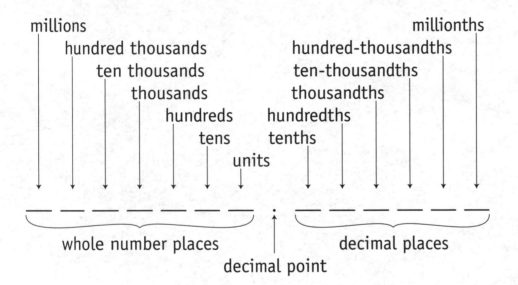

EXAMPLES The 6 in 2,658 has a value of 600.

The 8 in 183,492 has a value of 80,000.

The 4 in 2.54 has a value of $\frac{4}{100}$.

The 3 in $1.39 has a value of $\frac{3}{10}$ of a dollar or 30 cents.

The 5 in $1.85 has a value of $\frac{5}{100}$ of a dollar or 5 cents.

The 9 in $1.589 has a value of $\frac{9}{1,000}$ of a dollar.

Whole Number Operations

The four basic operations in arithmetic are addition, subtraction, multiplication, and division. The author and editor assume that readers of this book have a basic understanding of these four operations with whole numbers.

Addition

The answer to an addition problem is called the **sum** or the **total.** The words *combined* and *all together* suggest addition. You can add the numbers in an addition problem in any order.

EXAMPLE On Tuesday Lori wrote a check for $286, bought gasoline for $9, and spent $25 on a pair of jeans. All together, how much did she spend that day? Use a calculator to solve the problem.

Laurie spent $286 + $9 + $25 = $320 on Tuesday.

To solve the problem on a calculator, press:

The calculator display should read: 320.

Subtraction

The answer to a subtraction problem is called the **difference.**
Phrases like *how much more, how much less,* and *what is the balance due*
suggest subtraction. To solve a subtraction problem, put the larger of
two numbers on top. Or remember to enter the larger number first on
a calculator.

 Sam makes $36,300 a year, and his wife, Miriam, makes
$39,250. Miriam makes how much more than Sam in a year?

$39,250 − $36,300 = $2,950

A calculator has no key for the dollar sign ($) and no key for
the comma (,). To solve the example on a calculator, press:

Multiplication

The answer to a multiplication problem is called the **product.** You
will probably not see the word *product* in multiplication word problems.
A problem may tell you the rate for one thing and ask you to apply that
rate to several things. For example, a problem may give the price of one
item and ask you to find the price of several items. Or a problem may give
you the distance a car can travel on one gallon of gasoline and ask you to
find the distance the car can travel on several gallons.

EXAMPLE Cheryl pays $495 a month for rent. How much rent does she
pay in a year?

Multiply the monthly rent by the 12 months in a year.
12 × $495 = $5,940

To solve the example on a calculator, press:

Division

The answer to a division problem is called the **quotient.** You will probably not see the word *quotient* in word problems. A problem may give you information for several things and ask you to calculate the information for one thing. For example, a problem may tell you the price for three pounds of nails and ask you to calculate the price for one pound. Or a problem may ask you to find a *rate* such as miles per hour, miles per gallon, or dollars per hour. The word *per* suggests division.

To solve a division problem on a calculator, be sure to enter the number that is being divided into parts first.

> **EXAMPLE**   Jason makes $805 a week. If he works 35 hours each week, how much does he make per hour?
>
> **Divide his pay by the number of hours.**
>
> $805 ÷ 35 = $23 per hour
>
> **Notice that the $805 came first in the solution.**
>
> **To solve the example on a calculator, press:**
>
>

Rounding Whole Numbers

A *round* number is an approximation. For example, a salary of $27,934 is about $28,000. The word *round* refers to the zeros.

To round a whole number:

- Mark the digit in the place you are rounding to.
- If the digit to the right is greater than or equal to 5, add 1 to the marked digit.
- If the digit to the right is less than 5, leave the digit as it is.
- Change the digits to the right of the marked digit to zeros.

983 rounded to the nearest ten is 980.
6,154 rounded to the nearest hundred is 6,200.
$7,296 rounded to the nearest ten is $7,300.

In the last example, 9 is in the tens place: 9 + 1 = 10. Add the 1 in 10 to the 2 in the hundreds place. If you round $7,296 to the nearest hundred, you also get $7,300.

Estimation

An estimate is not exact. To estimate an answer, first round one of the numbers in the problem to the left-most digit.

  Sally pays $412 a month for rent. About how much rent does she pay in a year?

To the nearest hundred, Sally spends $400 a month for rent. She spends about 12 × $400 = $4,800 on rent in a year.

Finding an Average (or Mean) and a Median

An **average** or a **mean** is a number that represents a set of numbers. To find the average of a set of numbers, add the numbers. Then divide by the number of numbers in the set.

  Three houses were sold on Mrs. Jones' street last year. One sold for $84,600, another sold for $107,000, and a third sold for $94,300. What was the average selling price of the houses?

Add the three prices.

$84,600 + $107,000 + $94,300 = $285,900

Divide the total by 3: $285,900 ÷ 3 = $95,300

A **median** is another way to represent a set of numbers. The median is the middle value of a set of numbers.

EXAMPLE What was the median selling price of the three houses in the last example?

Arrange the prices in order from smallest to largest:

$84,600, $94,300, $107,000.

The median is the middle value, **$94,300.**

When two numbers are in the middle of a set of numbers, the median is the average of the two middle numbers.

EXAMPLE Over the weekend Sean looked at four used cars that are for sale. Their prices are $2,998, $3,490, $3,288, and $4,189. What is the median price of the cars?

Arrange the prices in order from smallest to largest.

$2,998, $3,288, $3,490, $4,189.

Find the average of the two middle values.

$3,288 and $3,490

$3,288 + $3,490 = $6,778

$6,778 ÷ 2 = $3,389

Decimals

A decimal is a part of a whole. A decimal is a kind of fraction. The whole is divided into 10 parts, 100 parts, 1,000 parts, and so on. The decimal places in our number system are to the right of the decimal point. The first six decimal places are listed on the chart on page 224.

Rounding Decimals

To round a decimal:

- Mark the digit in the place you are rounding to.
- If the digit to the right is greater than or equal to 5, add 1 to the marked digit.
- If the digit to the right is less than 5, leave the digit as it is.
- Drop the digits to the right of the marked digit.

3.49 rounded to the nearest tenth is 3.5.
0.1628 rounded to the nearest hundredth is 0.16.
7.0395 rounded to the nearest thousandth is 7.040.
$19.65 rounded to the nearest dollar is $20.
$1.449 rounded to the nearest cent is $1.45.

Adding Decimals

To add decimals, line up the numbers decimal point under decimal point. Remember that any whole number is understood to have a decimal point at the right.

EXAMPLE   Jane drove 2.8 miles to the grocery and 13 miles to her job. What total distance did she drive?

```
    2.8 miles
+ 13.
   15.8 miles
```

To solve the example on a calculator, press:

Subtracting Decimals

To subtract decimals, put the larger number on top and line up the decimal points. Use zeros to give each decimal the same number of decimal places.

EXAMPLE   The distance from Ann's house to her school is 8 miles. On her way to school she stopped at a sandwich shop that is 2.4 miles from her house. How much farther did she have to drive to get to her school?

```
    8.0 miles
 − 2.4
    5.6 miles
```

On a calculator, remember to enter the larger number in a subtraction problem first. To solve the example on a calculator, press:

Multiplying Decimals

To multiply decimals, multiply the numbers as you would multiply whole numbers. Then count the number of decimal places in the numbers you are multiplying. Put the total number of decimal places in the answer.

EXAMPLE Steve bought a package of cheese that weighed 0.85 pound. The price of the cheese was $6.95 per pound. How much did the package of cheese cost?

$ 6.95 two decimal places
✕ 0.85 two decimal places
 3475
 5560
$5.9075 four decimal places

The cheese cost $5.91.

To solve the example on a calculator, press:

Since our money system has only two decimal places (dimes and pennies), round the answer to the nearest cent (hundredths): $5.9075 rounds to $5.91.

Dividing Decimals

In the division problem 12 ÷ 4 = 3, the number 12 is the **dividend,** 4 is the **divisor,** and 3 is the **quotient.** To divide by a decimal, change the divisor to a whole number by moving the decimal point to the right. Then move the decimal point in the dividend the same number of places that you moved the point in the divisor. Finally, divide and bring the point up in the quotient above its new position in the dividend.

EXAMPLES

$$0.6\overline{)0.24} \text{ becomes } 6.\overline{)2.4}^{\,0.4}$$

Each decimal point moves one place.

$$0.003\overline{)2.7} \text{ becomes } 3.\overline{)2{,}700}^{\,900}$$

Each decimal point moves three places.

To solve the last example on a calculator, press:

Calculators take care of the decimal point for you. But remember to enter the dividend (the number that is being divided into parts) first.

EXAMPLE

Mike drove 100 miles on 4.5 gallons of gasoline. Calculate his gas mileage in miles per gallon to the nearest tenth.

100 ÷ 4.5 = 22.22… which rounds to 22.2 miles per gallon

On a calculator, press:

The answer on the calculator display will look something like 22.2222222. The calculator does not round the answer.

Fractions

A fraction expresses a part of a whole. The **denominator,** the bottom number in a fraction, tells the number of parts that the whole is divided into. The **numerator,** the top number, tells how many parts you have. For example, $\frac{3}{10}$ means that a whole is divided into ten parts, and you have three of them.

Reducing Fractions

The fractions $\frac{4}{8}$, $\frac{5}{10}$, and $\frac{17}{34}$ are equal. Each fraction reduces to $\frac{1}{2}$. To reduce a fraction, divide both the numerator and the denominator by a number that divides into each equally. Remember that reducing a fraction does not change its value.

EXAMPLES $\frac{45}{50} = \frac{9}{10}$

Divide both the numerator and the denominator by 5.

$\frac{70}{80} = \frac{7}{8}$

Divide both the numerator and the denominator by 10.

$\frac{75}{100} = \frac{3}{4}$

Divide both the numerator and the denominator by 25.

After you reduce a fraction, check to see whether another number divides evenly into both the numerator and the denominator.

EXAMPLE Reduce $\frac{48}{64}$.

$\frac{48}{64} = \frac{6}{8}$

Divide both the numerator and the denominator by 8.

$\frac{6}{8} = \frac{3}{4}$

Divide both the numerator and the denominator by 2.

Fractions and Decimals

A fraction is a kind of division problem. The fraction $\frac{1}{2}$ is a way of expressing the problem 1 divided by 2. To change a fraction to a decimal, divide the numerator by the denominator.

EXAMPLE Express $\frac{1}{2}$ as a decimal.

On a calculator, press:

The screen should read 0.5

The decimal form of $\frac{1}{2}$ is 0.5.

EXAMPLE Express $\frac{2}{3}$ as a decimal.

On a calculator, press: [2] [÷] [3] [=]

The screen should read 0.666666666

To the nearest hundredth, $\frac{2}{3}$ is equal to 0.67.

To change a decimal to a fraction, write the digits in the decimal as the numerator. The denominator is 10 if the decimal has one place, 100 if the decimal has two places, or 1,000 if the decimal has three places.

EXAMPLE Change 0.125 to a fraction.

$0.125 = \frac{125}{1,000} = \frac{5}{40} = \frac{1}{8}$

Finding What Part One Number Is of Another

In mathematics there are two common ways to compare numbers. One way is to subtract. For example, if Allen makes $15 an hour and Susan makes $18 an hour, we can say that Susan makes $3 an hour more than Allen ($18 − $15 = $3).

Another way to compare numbers is to make a fraction. In this kind of comparison, find what part one number is of another or what fraction one number is of another. The fraction represents a *part* over a *whole*.

EXAMPLE Julia makes $1,400 a month and pays $350 a month for rent. What part of her income does she spend on rent?

Make a fraction with the rent over the income.
Then reduce. $\frac{\$350}{\$1,400} = \frac{1}{4}$

Julia spends $\frac{1}{4}$ of her income on rent.

Basic Operations with Fractions

This book emphasizes the use of the calculator. Most calculators do not express fractions. The two examples below illustrate problems that can be solved as fraction problems or as decimal problems. (For more practice with fractions, look at *Number Power Fractions Percents and Decimals* in this series.)

EXAMPLE   What is the price of $4\frac{1}{2}$ pounds of meat that costs $4.80 a pound?

$$4\frac{1}{2} \times \$4.80 = \frac{9}{2} \times \frac{\$4.80}{1} = \frac{\$43.20}{2} = \$21.60$$

You can also change $4\frac{1}{2}$ to 4.5.

$$4.5 \times \$4.80 = \$21.60$$

To solve the example on a calculator, press:

EXAMPLE   Tom's Taxi Service charges $1.50 for the first $\frac{1}{8}$ mile and 25¢ for each additional $\frac{1}{8}$ mile. What is the total cost of a taxi ride that is $3\frac{1}{2}$ miles long?

First find the number of $\frac{1}{8}$-mile segments in $3\frac{1}{2}$ miles.

$$3\frac{1}{2} \div \frac{1}{8} = \frac{7}{2} \times \frac{8}{1} = \frac{56}{2} = 28 \text{ segments}$$

The first segment costs $1.50.
The next 27 segments cost 27 × $0.25 = $6.75.
The total cost is $1.50 + $6.75 = $8.25.

To solve the first part of the last example on a calculator, change $\frac{1}{8}$ and $3\frac{1}{2}$ to decimals. $\frac{1}{8}$ = 0.125 and $3\frac{1}{2}$ = 3.5.

To find the number of $\frac{1}{8}$-mile segments in $3\frac{1}{2}$ miles, press:

The reading on the screen should be: 28.

Percent

A **percent,** like a decimal or a fraction, is a part of a whole. The word *percent* means "out of 100." Percent is like a two-place decimal (hundredths).

To change a percent to a decimal, move the decimal point in the percent two places to the left. When there is no decimal point in a number, it is understood that there is a decimal point at the end.

  25% = 0.25
120% = 1.2
1.4% = 0.014
12.5% = 0.125

To change a decimal to a percent, move the decimal point two places to the right.

  0.5 = 50%
0.169 = 16.9%
3.25 = 325%
0.08 = 8%

Finding a Percent of a Number

The mathematical operation that is repeated most often in this book is finding a percent of a number. In these problems, the word *of* suggests multiplication.

EXAMPLE Matt's gross weekly income is $852. His employer deducts 20% of his wages for taxes and Social Security. What is Matt's net income each week?

Find 20% of $852. 20% as a decimal is 0.2.

Matt's employer deducts 0.2 × $852 = $170.40 from his weekly pay.

Matt's net weekly pay is $852 − $170.40 = $681.60.

There is more than one way to solve the last problem on a calculator.

On a calculator, press: [8] [5] [2] [×] [.] [2] [=]

Or press: [8] [5] [2] [×] [2] [0] [%]

The screen in each case should read 170.4 which represents $170.40, the amount the employer deducts. You still have to subtract $170.40 from Matt's gross pay to get his net pay.

Notice that you do not have to press the = key if you use the % key. Notice also that you may enter .2 first if you use the decimal method. If you use the % key, you must enter 852 first.

There is still another way to solve the last example. Matt's weekly income represents 100% of his income. Since the deductions are 20%, the net amount is:

$$100\% - 20\% = 80\% \text{ of his weekly income}$$

On a calculator, press: [8] [5] [2] [×] [.] [8] [=]

Or press: [8] [5] [2] [×] [8] [0] [%]

The screen in each case should read 681.6 which represents $681.60.

Look at one more example of finding a percent of a number.

EXAMPLE   Ruth bought a jacket for $49. What is the price of the jacket including 6% sales tax?

Find 6% of $49. $0.06 \times \$49 = \$2.94$

The price including sales tax is $49 + $2.94 = $51.94.

Think about the last example another way. If $49 is 100% of the price of the jacket, the price including sales tax is:

$$100\% + 6\% = 106\%$$

Find 106% of $49. $1.06 \times \$49 = \$51.94$

Finding What Percent One Number Is of Another

Problems in this book sometimes ask you to find what percent one number is of another. This is similar to finding what fraction one number is of another or what part one number is of another.

To solve these problems, first make a fraction with the part over the whole. Then reduce the fraction and change it to a percent.

EXAMPLE   Joanne takes home $2,400 a month. She spends about $720 each month on food. What percent of her take-home pay goes for food?

Make a fraction with the amount she spends on food over her whole income.

$$\frac{\$720}{\$2,400} = \frac{72}{240} = \frac{9}{30} = \frac{3}{10} = 30\%$$

Remember that a fraction is a kind of division problem. For the last problem, divide $720 by $2,400.

On a calculator, press:

The reading on the screen should be 0.3, which is equal to 30%.

Interest

Interest is a percentage of an amount of money. The formula for calculating interest is:

interest = principal × rate × time

where *principal* is the amount of money, *rate* is the percent, and *time* is the period, usually measured in years. The abbreviation for the formula is *i = prt*.

If the time is less than one year, change the time to a fraction of a year. For example:

6 months $= \frac{6}{12} = \frac{1}{2}$ or 0.5 year

9 months $= \frac{9}{12} = \frac{3}{4}$ or 0.75 year

1 year and 8 months $= 1\frac{8}{12} = 1\frac{2}{3}$ or 1.67 years

EXAMPLE Tomas borrowed $1,200 from his brother at 8% annual interest for one year and six months. How much interest did Tomas pay his brother?

Substitute $1,200 for the principal, 8% for the rate, and 1 year and 6 months for the time in the formula $i = prt$.

The rate of 8% = 0.08, and the time is $1\frac{6}{12} = 1\frac{1}{2}$ or 1.5 years.

$interest = \$1,200 \times 0.08 \times 1.5 = \$144$

To solve the example on a calculator, press:

(1)(2)(0)(0)(×)(·)(0)(8)(×)(1)(·)(5)(=)

EXAMPLE Altogether, how much does Tomas, in the last example, owe his brother at the end of one year and six months?

Tomas owes his brother the principal plus the interest.

$\$1,200 + \$144 = \$1,344$

Measurement

In the United States the system for measuring distance, weight, and liquid capacity is called the customary or standard system. Most of the rest of the world uses the metric system. On the next page is a chart with the customary units of measure.

You may have to change from one unit of measurement to an equivalent unit.

Customary Units of Measure	
Measures of Length 1 foot (ft)　　= 12 inches (in.) 1 yard (yd)　　= 36 inches 1 yard　　　　= 3 feet 1 mile (mi)　　= 5,280 feet 1 mile　　　　= 1,760 yards	**Liquid Measures** 1 pint (pt)　　= 16 ounces (oz) 1 cup　　　　= 8 ounces 1 pint　　　　= 2 cups 1 quart (qt)　　= 2 pints 1 gallon (gal)　= 4 quarts
Measures of Weight 1 pound (lb)　　= 16 ounces (oz) 1 ton (T)　　　= 2000 pounds	**Measures of Time** 1 minute (min) = 60 seconds (sec) 1 hour (hr)　　= 60 minutes 1 day　　　　= 24 hours

EXAMPLE　A room is 25 feet long. What is the length of the room in yards?

Divide the length by 3, the number of feet in 1 yard.

$25 \div 3 = 8\frac{1}{3}$ or about 8.3 yards

To read a time schedule, think of a 24-hour clock. 3:40 P.M. is the same as 12:00 noon + 3:40 = 15:40 hours.

EXAMPLE　Kathy took a train that left Pennsylvania Station at 11:45 in the morning. The train arrived in Albany at 2:10 in the afternoon. What was the total time of the ride?

The arrival time of 2:10 = 12:00 + 2:10 = 14:10.

Subtract the departure time from the arrival time.

$$\begin{array}{r} 14{:}10 \\ -\ 11{:}45 \\ \hline \end{array}$$

To subtract 45 from 10, change 14 hours to 13 hours and 60 minutes.

Then add 60 minutes to 10 minutes.

$$\begin{array}{r} 14{:}10 = 13 \text{ hr } 70 \text{ min} \\ -\ 11{:}45 = 11 \text{ hr } 45 \text{ min} \\ \hline 2 \text{ hr } 25 \text{ min} = \text{total travel time} \end{array}$$

Perimeter and Area

Perimeter is a measure of the distance around a flat figure. Perimeter is measured in inches, feet, yards, meters, and so on.

In this book you will need to calculate the perimeter of a rectangle. A rectangle is a four-sided figure with four right angles. The sides opposite each other are equal. The longer side is called the length, and the shorter side is called the width.

To find the perimeter of a rectangle, you can add the measurements of each of the four sides or you can use the formula $P = 2l + 2w$ where P is the perimeter, l is the length of the rectangle, and w is the width of the rectangle.

> **EXAMPLE**   What is the perimeter of a picture frame that is 20 inches long and 16 inches wide?
>
> Substitute 20 inches for the length and 16 inches for the width in the formula for the area of a rectangle.
>
> $P = 2l + 2w = 2(20) + 2(16) = 40 + 32 = 72$ inches

Area is a measurement of the amount of surface on a flat figure. Area is measured in square units such as square inches, square feet, and square yards.

In this book you need to calculate the area of a rectangle and the area of a triangle. The formula for the area of a rectangle is $A = lw$ where A is the area, l is the length, and w is the width.

Find the area of a floor that is 18 feet long and 11 feet wide.

Substitute 18 feet for the length and 11 feet for the width in the formula for the area of a rectangle.

$A = lw = 18 \times 11 = 198$ square feet

A triangle is a flat figure with three sides. The base of the triangle is the side on which the triangle appears to rest, and the height is the distance from the base to the highest point of the triangle. The formula for the area of a triangle is $A = \frac{1}{2}bh$ where A is the area, b is the base, and h is the height.

Find the area of a triangle with a base of 20 feet and a height of 12 feet.

Substitute 20 feet for the base and 12 feet for the height in the formula for the area of a triangle.

$A = \frac{1}{2}bh = \frac{1}{2} \times 20 \times 12 = 120$ square feet

Posttest 1

This test gives you a chance to check your understanding of the material in this book. When you finish the test, check your answers and review any lessons on which you need more work. You may use a calculator to solve these problems.

1. Tom earns $17.50 an hour for a 35-hour week. What is Tom's gross pay for a week?

2. The employer of Tom, in the last problem, deducts 7.65% of Tom's gross pay for F.I.C.A. How much is deducted from Tom's weekly paycheck for F.I.C.A.?

3. Use the following information to bring Sally's checkbook balance up-to-date. At the end of July, Sally's checkbook balance was $824.62. Her bank statement for July had a balance of $606.73. The statement included an interest payment of $0.48 and a service charge of $2.00. The statement did not include the following items: a check for $76.33 to her car mechanic, an ATM withdrawal of $140.00, and a deposit of $432.70.

4. Use the information in problem 3 to bring Sally's July bank statement up-to-date.

5. Calvin bought a shirt for $34.99 and a pair of jeans for $28.99. What was the total price of the items including sales tax of 4.5%?

6. A winter jacket that originally sold for $169 was on sale for 20% off the list price. What is the sale price of the jacket including a 6.5% sales tax?

7. Millie bought a package of cheese that weighed 0.85 pound and cost $7.22. To the nearest penny, what was the price of the cheese per pound?

8. A box of high-fiber cereal weighs 11.5 ounces and costs $2.99. What is the cost of the cereal per ounce?

9. Sarah and Frank went out to dinner. Frank ordered the roast beef dinner for $19.90, and Sarah ordered the chicken dinner for $14.90. They each had iced tea that cost $1.90 a glass, and they shared a piece of pie that cost $2.45. What was the total cost of the meal including 6% sales tax?

10. Frank, in the last problem, left a tip of $7.00. The tip was what percent of the cost of the meal?

11. A taxi company charges an initial fee of $2.00. Then, the ride costs $1.50 per mile, and there is a $2.00 charge for every 5 minutes of waiting time. What was the cost, including a 15% tip, of a $7\frac{1}{2}$-mile taxi ride that included a ten-minute delay for street construction?

12. A train left Memphis at 9:50 A.M. and arrived in New Orleans at 3:40 P.M. What was the travel time from Memphis to New Orleans?

13. To keep rabbits away from her vegetables, Laura wants to enclose a garden that is 36 feet long and 10 feet wide. The fencing she plans to use costs $1.39 per linear foot. Find the cost of the fencing that she needs to enclose the garden.

14. Bill and Sonia's living room is 20 feet long and 14 feet wide. Their dining room is 14 feet long and 10 feet wide. They want to buy wall-to-wall carpet that costs $25.90 a square yard. Find the cost of the carpet required to cover both the living room floor and the dining room floor.

15. Irene bought 200 shares of Fixit stock when the stock traded at 15.3 a share. She sold the shares two years later when the stock was trading at 19.25. Not counting commissions, what net profit did Irene make on the Fixit stock?

16. The broker for Irene, in the last problem, charged a commission of $20 plus 0.004 times the gross amount of the sale. Calculate the commission that Irene paid in order to sell her Fixit stock.

17. Bob borrowed $2,000 from his brother at 8% simple annual interest. When Bob repaid the loan, he paid his brother $2,240. For how long did Bob borrow the money?

18. Find the simple interest on $1,500 at 18% annual interest for one month.

19. Felix borrowed $7,500 at 8.5% simple annual interest for two years. What total amount of interest will he have to pay?

20. Zorah borrowed $5,000 at 16.5% simple annual interest for three years. If she pays the loan back in equal monthly installments, how much will she have to pay back each month?

21. Morris agreed to pay $200 down and $45 a month for 18 months in order to buy a wide-screen television. What total amount will he pay for the television?

22. The wide-screen television in the last problem is listed at $799. To the nearest tenth of a percent, what interest rate will Morris pay by buying the television in installments?

23. Casey's credit card company charges her an interest rate of 19.5% per annum. Find the monthly interest charge on an unpaid balance of $2,736.

24. Casey, in the last problem, must pay a minimum of $20 or 2.5% of the balance due. What is the minimum payment for her unpaid balance?

The Smiths have a 15-year mortgage of $85,000 at 7.5% annual interest. Their monthly payment is $880.67. Use this information to answer problems 25 and 26.

25. How much of the Smiths' first monthly payment is for interest?

26. How much of the Smiths' first monthly payment goes toward paying off the principal?

27. The reading on Gianna's electric meter last month was 37426 kwh. The reading this month is 37709 kwh. The company that provides her electricity charges 14.8314¢ per kilowatt-hour. Find the cost of the electricity that she used for the month.

28. Phil's gas company charges $0.312 per ccf of gas and a distribution charge of $0.126 per ccf of gas. The bill also includes a monthly fee of $6.00 and state tax of 4.5%. The reading on Phil's gas meter last month was 3251 ccf. The reading this month is 3287. Find the total of Phil's gas bill for the month.

29. Barbara does not want to spend more than 30% of her gross income for rent. Before taxes, she makes $37,500 a year. Find the maximum monthly rent that Barbara can pay if she sticks to her guideline.

30. The Douglas family bought their house in 1989 for $88,000. They sold the house in 2003 for $129,000. By what percent, to the nearest tenth, did the value of their house appreciate?

31. On January 1, the odometer (mileage gauge) in Steve's car had a reading of 62,412 miles. The reading on the last day of December was 74,650 miles. According to his receipts, Steve bought a total of 486 gallons of gasoline during the year. To the nearest tenth, what was the fuel efficiency in miles per gallon of his car?

32. Pam paid $19,500 for her car. Three years later her car was worth only $6,600. By what percent did the value of the car depreciate?

33. The homeowner's insurance for Paul and Maureen's house is 0.68% of the market value of the house. If their house has a market value of $142,900, what is the yearly premium for their insurance?

34. Yoshiko spends $3,241 a year for health insurance for herself and her three children. She received a notice that the premium will rise 5.5% next year. How much will the insurance cost Yoshiko next year?

35. The excise tax on airfare is $3.00 for each segment of a flight plus 7.5% of the price of the ticket. Find the excise tax on a $407 flight from New York to Houston with a stopover in Cincinnati.

36. The school taxes on the Jackson family's summer cabin are $6.853 per $1,000 of value. If their cabin is assessed at $99,500, what is their yearly bill for school taxes?

37. Phil and Carmen are married and filing their federal taxes jointly. Their combined taxable income for the year was $46,528. Use the following information from the 2003 tax tables to calculate the amount of tax they owe:

> For married taxpayers filing jointly, if taxable income is over $14,000 but not over $56,800, the tax is $1,400.00 plus 15% of the excess over $14,000.

38. To pay federal income tax, each month a total of $608 was withheld from the paychecks of Phil and Carmen, in the last problem. Calculate the additional amount of tax they owe or the amount of their refund.

39. Carolyn spends $638 a month for rent. Her average bill for gas and electricity each month is $47.50, and her average monthly telephone bill is $66.33. Her yearly take-home income is $31,095. What percent of her take-home income goes toward housing expenses?

40. Mr. and Mrs. Garcia expect to spend $132 a day for meals and $136 a day for lodging when they take their two children to Florida for vacation. How much should they save each month for twelve months to have enough cash for meals and lodging for a 10-day vacation in Florida?

Posttest 1 Evaluation Chart

Circle the number of any problem you missed. The column after the problem number tells you the lesson number where the skill is taught. The next column tells you the pages to review in this book.

Review the lessons that correspond to any of the problems that you got wrong.

Problem Number	Lesson Number	Review Pages	Problem Number	Lesson Number	Review Pages
1	1	11-20	21	11	111-118
2	1	11-20	22	11	111-118
3	2	21-30	23	12	119-128
4	2	21-30	24	12	119-128
5	3	31-40	25	13	129-138
6	3	31-40	26	13	129-138
7	4	41-48	27	14	139-152
8	4	41-48	28	14	139-152
9	5	49-56	29	15	153-162
10	5	49-56	30	15	153-162
11	6	57-68	31	16	163-172
12	6	57-68	32	16	163-172
13	7	69-78	33	17	173-186
14	7	69-78	34	17	173-186
15	8	79-90	35	18	187-196
16	8	79-90	36	18	187-196
17	9	93-100	37	19	197-209
18	9	93-100	38	19	197-209
19	10	101-110	39	20	212-221
20	10	101-110	40	20	212-221

Posttest 2: Write a Budget

This exercise gives you a chance to work on a budget for a family. The budget is for a family of four—David and Lena Ross, who are both 35 years old, and their two children, Catherine, who is seven, and Alex, who is five. David works full time as a landscaper for an urban park system. Lena works part time as a nurse in a medical clinic.

The family's expenses are separated into eight categories. Calculate a monthly subtotal for each category. For some expenses you will have to use a yearly cost to find the average monthly expense. For a few weekly expenses, simply multiply the cost by four to estimate a monthly total. For expenses that vary from month to month, you will see the monthly costs for September, October, and November. Find the average of these expenses to estimate the monthly expense.

When you finish, add the subtotals to find the family's total monthly expenses. Then calculate the percent of the entire budget that each category represents.

A. HOUSING

David and Lena own their home. They have a 30-year mortgage for $100,000 at a 7% annual interest rate. Look up their monthly mortgage payment in the table on page 132. They pay $80 each quarter for water and sewer. They are on a level-billing plan with the utility company that provides gas and electricity, and they pay $43 a month for these utilities. Their telephone bill in September was $78.56; for October, $49.23; and for November, $82.05. For trash removal they pay $4 a week. Their house has a market value of $147,500. Their community charges a property tax rate of $11.94 per $1,000 of market value. Their homeowner's insurance policy costs 0.69% of market value each year. Their only other housing expense for the year was a new roof on the garage, which cost $1,050.

Find the subtotal of their monthly housing expenses.

B. TRANSPORTATION

David drives a 12-year-old van that they already own. Lena drives a nearly new car for which they borrowed $5,500 at 8.5% annual interest for four years. Calculate the monthly payment on their car loan. (See Lesson 10.) The yearly insurance premium on the new car is $1,149, and on David's van the premium is $726. Each car gets a $25 oil change and lubrication twice a year. During the year David's van needed $500 in repairs. Together the vehicles consumed 808 gallons of gasoline in a year at an average price of $2.55 a gallon. The only other transportation cost for the family was $824 in airfare to visit Lena's parents in Canada.

Find the subtotal of their monthly transportation expenses.

C. FOOD

In September the family spent $338.28 on groceries and $127.42 in restaurants. In October they spent $350.82 on groceries and $88.78 in restaurants. In November, because of the Thanksgiving holiday, they spent $483.56 on groceries, but they spent only $42.89 in restaurants.

Find the subtotal of their average monthly food expenses.

D. CLOTHING

Their September credit card bills included charges of $84.72, $105.63, and $117.53 for back-to-school clothes and shoes for the children. In October they had only one clothing charge of $96.52. In November they had charges of $49.38 and $63.27 for clothing.

Find the subtotal of their average monthly clothing expenses.

E. HEALTHCARE

David's job includes health insurance, but the family has to make a co-payment each time they visit a doctor or go to the hospital. During the year they paid eight co-payments of $15 each for visits to the doctor and one co-payment of $240 when they took Catherine to the hospital. Each month David spends $39.76 on blood pressure medication, and all together the family spent another $100 during the year on medicine. Each of them goes to the dentist twice a year at a cost of $45 per visit. During the year, Lena spent $275 on an eye exam and new glasses, and David spent $195 for new sunglasses.

Find the monthly subtotal for their health care expenses.

F. ENTERTAINMENT

Every month David and Lena spend $21.95 for cable TV and $23.90 for Internet access. They subscribe to a weekly news magazine that costs $49.90 per year. David buys a newspaper for $0.75 twenty-one times a month, and they spend $3.50 four times a month for a Sunday paper. They spend about $16 a month on movies. Their other entertainment expenses for the year were a new television for $599 and a new bicycle for Catherine for $249.

Find the monthly subtotal for their expenses on entertainment.

G. PERSONAL INSURANCE & PENSIONS

David and Lena have both purchased 30-year life insurance policies. Find their monthly premiums on page 184. Together David and Lena paid $3,151.80 for pensions and Social Security through their employers.

Find the subtotal for their monthly expenses on personal insurance and pensions.

H. OTHER

In September David and Lena paid the total amount on their credit card bills. In October they left an unpaid balance of $100, and in November they left an unpaid balance of $250 on credit card bills. Every week they pay $30 for part-time child care for Alex. At Christmas, David and Lena usually spend about $400, and during the rest of the year they spend about $900 on gifts and charitable donations. One year from now, Lena would like to buy a new refrigerator that costs $699.

Find the monthly subtotal of these "other" expenses.

Check your answers on pages 278–279. Then round each subtotal to the nearest dollar, and fill in the summary budget below. Find the total of the eight categories, and calculate the percent that David and Lena spend on each category.

SUMMARY BUDGET

	Monthly Expense	Percent
A. HOUSING		
B. TRANSPORTATION		
C. FOOD		
D. CLOTHING		
E. HEALTHCARE		
F. ENTERTAINMENT		
G. PERSONAL INSURANCE & PENSIONS		
H. OTHER		
TOTAL		

Again, check your answers on pages 278–279. Then compare the Ross family's budget with the table of Average Household Expenditures on page 213. Finally, try to make a budget for yourself and your family.

ANSWER KEY

Pretest

Pages 1–7

1. 129,000

2. $44,000

3. $12,000 \div 436 = 27.5...$ or 28 mpg

4. $12 \times \$643 = \$7,716$

5. $\$4,599 + \$6,399 + \$5,469 = \$16,467$
   $\$16,467 \div 3 = \$5,489$

6. $\$106 + \$62 = \$168$
   $\$168 \div 2 = \$84$

7. 2.8

8. $39.57

9. $0.85 + 1.3 + 9.418 = 11.568$

10. $0.76 - 0.3 = 0.46$

11. $1.07 \times \$9.89 = \$10.5823$ or $10.58

12. $21 \div 3.5 = 6$

13. $2.15 \times \$7.99 = \$17.1785$ or $17.18

14. $\$887.43 - \$435 - \$58.27 - \$127.96 + \$301.68 = \$567.88$

15. 11.3 means $11.30 per share.
    $200 \times \$11.30 = \$2,260$

16. 17 mills = $0.017
    $\$115,000 \times \$0.017$ per dollar = $1,955.

17. $\frac{36}{54} = \frac{6}{9} = \frac{2}{3}$

18. $\frac{125}{275} = \frac{5}{11}$

19. $\frac{1}{5} = 0.2$
    $3 \div 0.2 = 15$ segments

20. $\frac{\$316}{\$1,580} = \frac{1}{5}$

21. $0.065 \times \$1,400 = \$91$

22. $1.08 \times \$72 = \$77.76$

23. $\frac{85}{130} = 0.6538...$ or 65.4%

24. $\frac{\$4,800}{\$16,000} = 0.3 = 30\%$

25. $1.15 \times \$38.70 = \$44.505$ or $44.51

26. $1.06 \times \$18.50 = \$19.61$

27. $0.8 \times \$129 = \$103.20$

28. $\frac{\$120}{\$2,000} = 0.06 = 6\%$

29. $\$18,000 - \$6,800 = \$11,200$
    $\frac{\$11,200}{\$18,000} = 0.622...$ or 62%

30. $\$90,000 - \$65,000 = \$25,000$
    $\frac{\$25,000}{\$65,000} = 0.384...$ or 38%

31. 3 months $= \frac{3}{12} = \frac{1}{4} = 0.25$ year
    interest = $\$2,800 \times 0.09 \times 0.25 = \$63$

32. interest = $\$4,000 \times 0.18 \times \frac{1}{12} = \$60$

33. 145 yards $\times$ 3 feet per yard = 435 feet

34. 2:12 = 14:12 or 14 hr 12 min
    14 hr 12 min = 13 hr 72 min
    $$\underline{-10 \text{ hr } 40 \text{ min}}$$
    3 hr 32 min

35. 2.5 pounds $\times$ 16 ounces per pound = 40 ounces

36. 1 gallon = 4 quarts
    $\frac{1}{2}$ gallon = 2 quarts
    $\$1.89 \div 2 = \$0.945$ or $0.95 per quart

37. $P = 2l + 2w = 2(40) + 2(24) = 80 + 48 = 128$ feet

38. $A = lw = 40 \times 24 = 960$ square feet

39. $l = 18 \div 3 = 6$ yards
    $w = 12 \div 3 = 4$ yards
    $A = lw = 6 \times 4 = 24$ square yards

40. $A = \frac{1}{2}bh = \frac{1}{2} \times 30 \times 14 = 210$ square feet

Lesson 1: Income

Pages 11–20

Pre-Lesson Vocabulary Practice

1. wage
2. deductions
3. yearly salary
4. withhold
5. overtime
6. income

Exercise 1

Part A

1. $40 \times \$12.80 = \$512$

2. $40 \times \$8.25 = \$330$
   $7 \times 1.5 \times \$8.25 = \$86.625$ or $86.63
   $\$330 + \$86.63 = \$416.63$

3. $44,720 \div 52 = \$860$

4. $0.05 \times \$21,518 = \$1,075.90$

5. $\$477.75 \div 35 = \$13.65$

6. $0.0765 \times 35 \times \$17.90 = \$47.927...$ or $\$47.93$

7. $68

8. $54

9. $.06 \times \$159,500 = \$9,570$

10. $0.035 \times \$16,583 = \$580.405$ or $\$580.41$
$0.054 \times \$34,792 = \$1,878.768$ or $\$1,878.77$
$\$580.41 + \$1,878.77 = \$2,459.18$

Part B

11. $\$42,680 \div 52 = \$820.769...$ or $\$820.77$

12. $0.0765 \times \$820.77 = \$62.788...$ or $\$62.79$

13. $103

14. deductions $= \$62.79 + \$103 + \$37.50 = \$203.29$
net income $= \$820.77 - \$203.29 = \$617.48$

Part C

15. $40 \times \$23.80 = \$952$

16. $0.0765 \times \$952 = \$72.828$ or $\$72.83$

17. $62

18. deductions $= \$72.83 + \$62 + \$12.50 = \$147.33$
net income $= \$952 - \$147.33 = \$804.67$

Part D

19. $\$52,380 \div 52 = \$1,007.307...$ or $\$1,007.31$

20. $0.0765 \times \$1,007.31 = \$77.059...$ or $\$77.06$

21. $87

22. deductions $= \$77.06 + \$87 + \$45 = \$209.06$
net income $= \$1,007.31 - \$209.06 = \$798.25$

Post-Lesson Vocabulary Reinforcement

1. By law

2. rate, $1\frac{1}{2}$ times

3. paycheck, union dues, charitable

4. withholding allowances

5. guidelines

6. find, earns, decimals

7. % of

8. e   **13.** h

9. f   **14.** d

10. c   **15.** b

11. i   **16.** j

12. g   **17.** a

Language Builder

Activity A

1. Assume, Divide, Look at, review, Use, Remember

Activity B

2. Multiply your hourly wage by the number of hours you work each week.

3. Assume that a year has 52 weeks.

4. Divide Sara's salary by 52.

5. Look at the first table.

Activity C

6. worked   **9.** claimed

7. contributed   **10.** sold

8. withheld

Activity D

11. How much did John make in overtime wages in a week when he worked 46 hours?

12. John claimed three withholding allowances on his W-4 form.

13. What was Eve's commission on a house that sold for $159,500?

Lesson 2: Banking

Pages 21–30

Exercise 2

Part A

1. $\$371.49 + \$0.26 - \$2.50 = \$369.25$

2. $\$357.97 + \$125 = \$482.97$

3. $\$60.00 + \$53.72 = \$113.72$

4. $\$482.97 - \$113.72 = \$369.25$

5. Yes. The corrected check register of $369.25 matches the corrected bank statement. The answers to questions 1 and 4 are the same, so the checkbook has been reconciled.

Part B

6. $763.28 + $0.27 – $3.25 = $760.30

7. $902.96 + $209.83 = $1,112.79

8. $140 + $80.00 + $19.53 + $112.96 = $352.49

NUMBER	DATE	DESCRIPTION OF TRANSACTION	PAYMENT/DEBIT(-)		DEPOSIT/CREDIT(+)		$	905	94
			$		$			19	53
112	4/23	check	19	53				886	41
	4/23	deposit			209	83		209	83
								1,096	24
113	4/24	check	112	96				112	96
								983	28
	4/27	ATM	140	–				140	–
								843	28
	4/28	ATM	80	–				80	–
								763	28
	4/29	interest			0	27		0	27
								763	55
	4/29	service charge	3	25				3	25
								760	30

Part C

9. $661.96 + $0.43 – $4.00 = $658.39

10. $193.41 + $614.96 = $808.37

11. $100 + $49.98 = $149.98

12. $808.37 – $149.98 = $658.39

13. Yes. The corrected check register of $658.39 equals the corrected bank statement.

Part D

14.

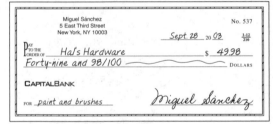

Post-Lesson Vocabulary Reinforcement

1. c   8. b
2. e   9. d
3. a   10. a
4. h   11. b
5. g   12. e
6. f   13. c
7. d

Language Builder

Activity A

1. Answer may vary, but should resemble the sentences below.

   a. To reconcile a checking account with a bank statement, first bring the checkbook register up to date by adding any interest earned and subtracting any service charges.

   b. To correct Keisha's April checkbook register, add the interest that she earned for the month and subtract the service charge.

   c. To bring Tim's checkbook records up to date, add the interest and subtract the service charge.

   d. To bring the bank statement up to date, add the late deposit, subtract checks that have not cleared, subtract the ATM withdrawal, and subtract the debit card purchase.

   e. To solve the problems in the next exercise, review adding and subtracting decimals, page 230.

   f. To correct the bank statement, first add any unrecorded deposits to the total on the statement.

Activity B

2. c   3. b   4. a   5. d

Activity C

6. ✓   7. no gerunds   8. ✓

Activity D

9. reconciling, adding, subtracting, subtracting

10. Balancing, adding, subtracting

Lesson 3: Shopping

Pages 31–40

Pre-Lesson Vocabulary Practice

1. e   4. d
2. a   5. f
3. b   6. c

Exercise 3

Part A

1. 1.8 × $1.99 = $3.582 or $3.58
   1.3 × $0.99 = $1.287 or $1.29
   $3.58 + $1.29 = $4.87

2. 2.4 × $0.89 = $2.136 or $2.14
   $2.14 + $1.69 + $3.19 = $7.02

3. 3.65 × $1.09 = $3.9785 or $3.98
   1.05($4.99) = $5.2395 or $5.24
   $3.98 + $5.24 = $9.22

4. $0.45 \times \$7.19 = \$3.2355$ or $\$3.24$
$2.1 \times \$1.49 = \$3.129$ or $\$3.13$
$\$3.24 + \$3.13 = \$6.37$

5. $2 \times \$1.49 = \$2.98$
$2 \times \$1 = \$2$
$\$1.79 - \$0.50 = \$1.29$
$\$2.98 + \$2.00 + \$1.29 = \$6.27$
$1.06 \times \$6.27 = \$6.65$

Part B

6. Arizona

7. Missouri

8. $1.06 \times \$29.98 = \$31.7788$ or $\$31.78$

9. $1.056 \times \$154.95 = \$163.6272$ or $\$163.63$

10. Food items are exempt from tax in California.
In Missouri, the tax is $0.0835 \times \$48.27 =$
$\$4.030545$ or $\$4.03$.
The total in Missouri is $\$4.03$ more expensive
than in California.

Part C

11. $0.1 \times \$399 = \$39.90$
$\$399.00 - \$39.90 = \$359.10$
$1.06 \times \$359.10 = \$380.646$ or $\$380.65$

12. 15% off means $100\% - 15\% = 85\%$ of the
list price

13. 40% off is 60% of the original.
$0.6 \times \$139 = \$83.40$
$1.075 \times \$83.40 = \$89.655$ or $\$89.66$

14. 10% off is 90% of the original.
$0.9 \times \$199 = \$179.10$
$1.0835 \times \$179.10 = \$194.05485$ or $\$194.05$

15. $\$199 - \$194 = \$5$

Part D

16. $\frac{\$50}{\$250} = \frac{1}{5} = 20\%$

17. $\frac{\$150}{\$2,000} = 7.5\%$

18. $\$299 - \$50 = \$249$

19. $\$2,460 - \$300 = \$2,160$
$1.05 \times \$2,160 = \$2,268$

Post-Lesson Vocabulary Reinforcement

1. consumer

2. Sales tax

3. discount

4. Sale price

5. coupon

6. exempt, price

7. add, on top of

8. end-of-season, off

9. save

10. c

11. a

12. b

13. e

14. d

Language Builder

Activity A

1. sold, known, underlined

Activity B

2. A discount means that the list price is reduced.

3. What is the sale price of a shirt that is listed at
$\$24.89$ less 20%?

4. At an end-of-the-season sale, winter coats are
marked 40% off.

5. A reclining chair is listed at $\$299$ at Phil's
Furniture.

6. A sofa and matching armchair are listed at
$\$1,999$ at Phil's.

Activity C

7. minimum   **8.** maximum

9. Which state has the <u>highest</u> tax rate?

10. Calculate the difference between the list price
and the sale price to the <u>nearest</u> dollar.

Lesson 4: Unit Pricing

Pages 41–48

Pre-Lesson Vocabulary Practice

1. gallon

2. quart

3. pint

4. ounce

Exercise 4

Part A

1. $\$4.13 \div 2.6 = \$1.588\ldots$ or about $\$1.59$ per pound

2. $\$12.49 \div 180 = \$0.069\ldots$ or about $\$0.07$ per tablet

3. $\frac{1}{2}$ gallon $= 2$ quarts
$\$1.79 \div 2 = \$0.895$ or about $\$0.90$ per quart

4. 1 gallon costs $2 \times \$1.79 = \$3.58$

5. $\$3.29 \div 100 = \$0.0329$ or about $\$0.03$ per
square foot

6. length = $66\frac{2}{3} \times 3$ feet per yard $= \frac{200}{3} \times \frac{3}{1} =$ 200 feet
width = 12 inches = 1 foot
Area = 200×1 = 200 square feet

7. Ace costs $2.79 \div 75 = \$0.0372$ per square foot. Baxter costs $0.99 \div 25 = \$0.0396$ per square foot. Ace is the better buy.

8. Brand A costs $1.99 \div 36 = \$0.05527...$ each. Brand B costs $5.00 \div 200 = \$0.025$ each. Brand C costs $2.69 \div 100 = \$0.0269$ each. Brand B is the best buy.

9. Healthy Start costs
  $3.29 \div 12.5 = \$0.263...$ per ounce.
Better Body costs
  $3.79 \div 11.4 = \$0.332...$ per ounce.
Good Grains costs
  $3.19 \div 10.75 = \$0.296...$ per ounce.
Healthy Start is the best buy.

Part B

10. West Coast

11. Rocky Mountains

12. East Coast

13.

Region	Total Cost			Unit Price
U.S. Average	$57.25	÷ 18.5 =		$3.095
East Coast	$57.45	÷ 18.5 =		$3.105
New England	$58.69	÷ 18.5 =		$3.172
Central Atlantic	$58.19	÷ 18.5 =		$3.145
Lower Atlantic	$56.40	÷ 18.5 =		$3.049
Midwest	$57.02	÷ 18.5 =		$3.082
Gulf Coast	$54.46	÷ 18.5 =		$2.944
Rocky Mountains	$53.47	÷ 18.5 =		$2.890
West Coast	$60.59	÷ 18.5 =		$3.275

14. $29.44 – $28.90 = $0.54

15. $3.388 – $3.095 = $0.293 or about $0.29

Post-Lesson Vocabulary Reinforcement

1. napkins
count

2. string beans
weight

3. shampoo
liquid

4. aluminum foil
area

5. wisely, compare

6. weight

7. unit price count

8. required, to display weight

9. unit

10. packaged area

11. better buy

Language Builder

Activity A

1. highest

2. higher

3. easier, quicker

4. nearest

5. best, greatest

Activity B

6. shop / wisely

7. find / quickly

8. Read / carefully

Activity C

Answers will vary.

Lesson 5: Dining Out

Pages 49–56

Pre-Lesson Vocabulary Practice

1. sum

2. sales tax

3. check or bill

4. tip or gratuity

5. amount tendered, amount received

Exercise 5

Part A

1. $11.95 + $1.15 + $1.85 + $9.95 + $1.15 + $1.55 = $27.60

2. $0.06 \times \$27.60 = \$1.656$ or $1.66

3. (4) $4.15
$0.15 \times \$27.60 = \$4.14.$
$4.15 is closest.

4. $27.60 + $1.66 + $4.25 = $33.51

Part B

5. $4 + 2($4 + $2) + $5 + $2 + $2 = $25

6. $3.75 + 2($4.25 + $1.95) + $5.25 + 2($0.75) + 2($1.25) = $25.40

7. $25.40 \div 4 = \$6.35

8. No. The pre-tax total is $25.40.
   6% sales tax = 0.06 × $25.40 = $1.524 or $1.52
   15% tip = 0.15 × $25.40 = $3.81
   The total is $25.40 + $1.52 + $3.81 = $30.73.
   Thirty dollars is not quite enough.

Part C

9. $27.90 ÷ 2 = $13.95

10. Choice (2) is the largest.
    (1) $3.749 or $3.75
    (2) 4 × $1.44 = $5.76
    (3) $3.605 or $3.61
        $3.61 ÷ 2 = $1.805 or $1.81
        $3.61 + $1.81 = $5.42
    (4) 3 × $1.44 = $4.32; $4.32 + $1.00 = $5.32

11. No. $40.00 − $37.49 = $2.51, which is not an adequate tip.

12. $37.49 + $5.50 = $42.99
    $42.99 ÷ 2 = $21.495 or $21.50

13. Choice (2) is the best method. If you triple the tax rate you get 3 × 5% = 15%. In other words, three times the sales tax is exactly 15% of the pre-tax amount.

Post-Lesson Vocabulary Reinforcement

1. tip or gratuity
2. adequate
3. before-tax total
4. exceptionally
5. leave
6. round
7. all together
8. cover
9. percentage
10. result
11. as well as
12. check or bill
13. g
14. f
15. d
16. b
17. a
18. c
19. e
20. h
21. i
22. l
23. k
24. j

Language Builder

Activity A

Answers will vary.

Activity B

1. have, ate
2. get, tendered
3. had, is
4. is, left
5. got, rounded

Activity C

6. Move the decimal point one place to the left.
7. Add the tax to the bill.
8. Solve the problems in all the exercises.
9. Pay all of the bill.
10. Triple the sales tax to calculate the tip.

Lesson 6: Travel

Pages 57–68

Pre-Lesson Vocabulary Practice

1. a
2. i
3. b
4. f
5. c
6. g
7. e
8. h
9. d

Exercise 6

Part A

1. $1\frac{1}{2} ÷ \frac{1}{6} = \frac{3}{2} × \frac{6}{1} = \frac{18}{2} = 9$ segments
   each $\frac{1}{6}$ mile long first segment cost = $1.80
   8 remaining segments cost = 8 × $0.30 = $2.40
   basic cost = $4.20

2. $2\frac{3}{4} ÷ \frac{1}{6} = \frac{11}{4} × \frac{6}{1} = \frac{66}{4} = 16\frac{1}{2}$ segments
   first segment cost = $1.80
   $15\frac{1}{2}$ remaining segments cost =
   15.5 × $0.30 = $4.65
   6 minute delay cost = 6 × $0.20 = $1.20
   basic cost = $1.80 + $4.65 + $1.20 = $7.65
   cost including 15% tip = 1.15 × $7.65 =
   $8.797... or $8.80
   A passenger might round the total to $9.00.

3. $\frac{1}{2} ÷ \frac{1}{6} = \frac{1}{2} × \frac{6}{1} = \frac{6}{2} = 3$ segments
   first segment cost = $1.80
   2 remaining segments cost = 2 × $0.30 = $0.60
   3-minute delay cost = 3 × $0.20 = $0.60
   basic cost = $1.80 + $0.60 + $0.60 = $3.00
   cost including 15% tip = 1.15 × $3.00 = $3.45
   A passenger might round the total to $3.50.

Part B

4. initial fare = $1.90
   2.5 × $1.60 = $4.00
   additional person = $0.50
   basic cost = $6.40

5. initial fare = $1.90

6.25 × $1.60 = $10.00

12-minute wait = 2 × $2.00 = $4.00

basic cost = $1.90 + $10.00 + $4.00 = $15.90

cost including tip = 1.15 × $15.90 = $18.285

A passenger might pay $18.25 or $18.00 or $18.50.

Part C

6. a.

$$7:03 = 6 \text{ hr } 63 \text{ min}$$
$$- 6:19 = 6 \text{ hr } 19 \text{ min}$$
$$\overline{\phantom{- 6:19 = 6 \text{ hr }} 44 \text{ min}}$$

b.

$$9:19 = 8 \text{ hr } 79 \text{ min}$$
$$- 8:40 = 8 \text{ hr } 40 \text{ min}$$
$$\overline{\phantom{- 8:40 = 8 \text{ hr }} 39 \text{ min}}$$

c.

$$6:29 = 5 \text{ hr } 89 \text{ min}$$
$$- 5:38 = 5 \text{ hr } 38 \text{ min}$$
$$\overline{\phantom{- 5:38 = 5 \text{ hr }} 51 \text{ min}}$$

d.

$$9:32 = 8 \text{ hr } 92 \text{ min}$$
$$- 8:58 = 8 \text{ hr } 58 \text{ min}$$
$$\overline{\phantom{- 8:58 = 8 \text{ hr }} 34 \text{ min}}$$

7. $76.00 ÷ 10 = $7.60

8. $80.00 ÷ 10 = $8.00

9. $48.45 ÷ 10 = $4.845 or $4.85

10. $51.00 ÷ 10 = $5.10

11. $171.50 ÷ 44 = $3.897...or $3.90

Part D

12. initial fare = $2.00

$5 ÷ \frac{1}{5} = \frac{5}{1} \times \frac{5}{1} = 25$ segments

25 × $0.30 = $7.50

basic cost = $2.00 + $7.50 = $9.50

cost including tip = 1.15 × $9.50 = $10.925 or $10.93

A passenger would probably pay $11.00.

13. initial fare = $2.00

$1\frac{1}{2} ÷ \frac{1}{5} = \frac{3}{2} \times \frac{5}{1} = \frac{15}{2} = 7\frac{1}{2}$ segments

7.5 × $0.30 = $2.25

delay cost = 8 × $0.20 = $1.60

night surcharge = $0.50

basic cost = $2.00 + $2.25 + $1.60 + $0.50 = $6.35

cost including tip = 1.15 × $6.35 = $7.3025 or $7.30

14. initial fare = $2.00

$10 ÷ \frac{1}{5} = \frac{10}{1} \times \frac{5}{1} = 50$ segments

50 × $0.30 = $15.00

delay cost = 15 × $0.20 = $3.00

night surcharge = $0.50

basic cost = $2.00 + $5.00 + $3.00 + $0.50 = $20.50

cost including tip = 1.15 × $20.50 = $23.575.

The passengers might round the total to $24.00.

Then each passenger owes $24.00 ÷ 3 = $8.00

Part E

15. From Chicago departure to midnight is

12 − 8 = 4 hours.

From midnight until arrival in New Orleans is

12 + 3:40 = 15 hr 40 min.

Total time = 19 hr 40 min

16. From New Orleans departure to midnight is

$$24:00 = 23 \text{ hr } 60 \text{ min}$$
$$- 13:55 = 13 \text{ hr } 55 \text{ min}$$
$$\overline{ 10 \text{ hr } 5 \text{ min}}$$

From midnight until arrival is 9 hours.

Total time = 19 hr 5 min

17. 6:50 − 6:27 = 23 min

18. total distance = 926 miles

$142 ÷ 926 = $0.1533... or $0.153 per mile

19. The train leaves Mattoon at 11:13 and arrives in Fulton at 3:14.

From 11:13 to noon is 47 minutes. From noon to 3:14 is 3 hours and 14 minutes.

Total time = 4 hr 1 min

20. The train leaves McComb at 3:48 and arrives in Memphis at 10:00.

10:00 − 3:48

Total time = 6 hr 12 min

21. The train arrives in Carbondale at 1:21 and departs at 1:26. Total layover = 5 min

22. Subtract: 873 − 254 = 619 miles

23. Subtract: 741 − 25 = 716 miles

Post-Lesson Vocabulary Reinforcement

1. train schedule

2. departure city

3. departure time

4. arrival time

5. arrival city

6. b

7. a

8. e

9. d

10. c

11. Public transportation

12. digital clock

13. analog clock

14. a layover

15. off-peak

Language Builder

Activity A

1. From the time the <u>meter</u> <u>is</u> <u>turned</u> on, a taxi ride in Philadelphia costs $1.80.

2. If a <u>taxi ride</u> <u>is</u> <u>stopped</u> for an accident, the waiting-time rate takes effect.

3. How much is a train ticket <u>that</u> <u>is</u> <u>purchased</u> on the Web?

4. Find the cost of two tickets <u>that</u> <u>are</u> <u>bought</u> at the station?

Activity B

5. are controlled

6. is charged, is delayed

7. are stopped

Activity C

8. if there are no delays

9. if there is no waiting time

10. if there are 8 minutes of waiting time

11. if they divide the fare and a 15% tip evenly

Lesson 7: Home Renovation

Pages 69–78

Pre-Lesson Vocabulary Practice

1. g
2. f
3. h
4. d
5. e
6. j
7. k
8. b
9. a
10. i
11. c

Exercise 7

Part A

1. "To go around" suggests perimeter.

2. Tiles cover an area.

3. Flooring covers an area.

4. "To make a border for" suggests perimeter.

5. "To cover the walls" suggests area.

6. "To apply to the boards on a deck" suggests area.

Part B

7. The base of the triangle is 20 ft, and the height is $16 - 9 = 7$ ft.
   The area of the triangle is $A = \frac{1}{2} \times 20 \times 7 = 70$ sq ft.
   The area of the rectangle is $A = 20 \times 9 = 180$ sq ft.
   The total area of the front wall is $70 + 180 = 250$ sq ft.

8. The side wall is 22 ft long and 9 ft high.
   The area of the side wall is $A = 22 \times 9 = 198$ sq ft.

9. The garage has two end walls and two side walls. The total surface area is
   $2(250) + 2(198) = 500 + 396 = 896$ sq ft.

10. $896 \div 350 = 2.56$ gallons of paint. Carl should buy 3 gallons.

11. $3 \times \$22.95 = \$68.85$

Part C

12. The perimeter of each frame is
    $P = 2\left(4\frac{1}{2}\right) + 2\left(2\frac{1}{2}\right) = 9 + 5 = 14$ ft.

    For two frames, Maureen needs $2 \times 14 = 28$ ft of molding.

13. Maureen bought $28 + 2 = 30$ feet of molding.
    $30 \times \$2.89 = \$86.70$
    $1.065 \times \$86.70 = \$92.3355$ or $92.34

14. $A = 14 \times 10.5 = 147$ sq ft
    $147 \div 9 = 16.33...$ or 16.3 square yards

15. $16.3 \times \$32.95 = \$537.085$ or $537.09
    $537.09 + \$15.00 = \$552.09

Part D

16. The triangle on the end wall has a base of 40 ft and a height of $38 - 24 = 14$ ft.
    The area of the triangle is
    $A = \frac{1}{2} \times 40 \times 14 = 280$ sq ft.

    The rectangle on the end wall is 40 ft long and 24 ft high.
    The area of the rectangle is $A = 40 \times 24 = 960$ sq ft.
    The area of the end wall is $280 + 960 = 1,240$ sq ft.
    The area of the side wall is $A = 56 \times 24 = 1,344$ sq ft.
    The total wall area is $2(1,240) + 2(1,344) = 2,480 + 2,688 = 5,168$ sq ft.

17. $5,168 \div 400 = 12.92$ or 13 gallons

18. $2 \times 13 \times \$19.85 = \$516.10$

19. $P = 2(65) + 2(5) = 130 + 10 = 140$ ft
    $140 \times \$0.89 = \$124.60$

20. The area of the new room is $A = 24 \times 18 = 432$ sq ft.
    The low estimate is $432 \times \$32 = \$13,824$.
    The high estimate is $432 \times \$54 = \$23,328$.
    The difference is $\$23,328 - \$13,824 = \$9,504$.

Post-Lesson Vocabulary Reinforcement

1. gallon
2. coverage
3. home renovation
4. Do-it-yourself projects
5. home owners, renters
6. to build
7. to raise the value of a property
8. to surround
9. find the difference
10. delivery charge
11. approximately
12. rectangle
13. width
14. length
15. triangle
16. height
17. base
18. foot
19. meter
20. yard

Language Builder

Activity A

1. What
2. How
3. How
4. What

Activity B

5. does
6. is
7. will

Activity C

8. <u>To raise</u> the value of a property, many owners use do-it-yourself projects.

9. What is the total cost of the fencing required <u>to surround</u> the vegetable garden?

10. How much paint is needed <u>to cover</u> the end wall of the shed?

Lesson 8: Investing

Pages 79–90

Pre-Lesson Vocabulary Practice

1. h	8. e	15. a
2. k	9. b	16. e
3. i	10. a	17. f
4. f	11. j	18. g
5. g	12. c	19. h
6. d	13. d	20. b
7. l	14. c	

Exercise 8

Part A

1. $40 \times \$38.40 = \$1,536$

2. $\$80 \div \$2,000 = 0.04 = 4\%$

3. $723.647 \times \$15.90 = \$11,505.987$ or $\$11,506$

Part B

4. gross amount $= 80 \times \$46.30 = \$3,704$
   $0.002 \times \$3,704 = \$7.408$ or $\$7.41$
   total cost $= \$3,704 + \$35 + \$7.41 = \$3,746.41$

5. The fee is $19.95 for all Internet transactions.

6. $20 \times \$32.45 = \$649$
   fee $= \$19.95$
   net amount $= \$649 - \$19.95 = \$629.05$

7. For the purchase:
   $60 \times \$22.70 = \$1,362$
   $0.002 \times \$1,362 = \$2.724$ or $\$2.72$
   total cost $= \$1,362 + \$35 + \$2.72 = \$1,399.72$

   For the sale:
   $60 \times \$26.30 = \$1,578$
   $0.002 \times \$1,578 = \$3.156$ or $\$3.16$
   total fee $= \$35 + \$3.16 = \$38.16$
   Sandra receives $\$1,578 - \$38.16 = \$1,539.84$

   Sandra's net income is
   $\$1,539.84 - \$1,399.72 = \$140.12$

Part C

8. Cadmium (CAD)
9. Branscom (BRA)
10. Romax (RMX)
11. Branscom (BRA)
12. 30.15
13. Romax (RMX)
14. $240 \times 100 = 24,000$

15. Branscom, NorthAir, and Romax were all down for the day.

16. Branscom. $7.75 is almost twice $4.00, the low price for the year.

17. $81.75 − $57.95 = $23.80

18. The stock was up $0.44 from the previous day's closing price.
    The previous day's closing price was $0.44 less than today's closing price.
    $9.92 − $0.44 = $9.48.

19. $30.15 − $27.80 = $2.35

Post-Lesson Vocabulary Reinforcement

1. f
2. g
3. c
4. a
5. b
6. d
7. h
8. e
9. n
10. i
11. l
12. o
13. k
14. m
15. j
16. traded
17. Index fund
18. treasury bonds
19. CDs
20. municipal bonds
21. corporate bonds
22. IRAs
23. net asset value

Language Builder

Activity A

1. Maria received $385 when she sold her shares of
     S   V   CM S V
   Curall.

2. The number 12.6 means that one share of Curall
        S   V   CM   S

   cost $12.60 when Maria bought it.
     V   CM   S   V

3. Stocks are usually traded through a stockbroker
     S   V

   who charges a commission.
   S = CM V

4. The listing is for a company called Marlex
        S V

   that   trades under the symbol MLX.
   S = CM   V

Activity B

Possible answers:

5. CDs are insured deposits that pay interest and require that the money remain invested for a fixed period of time.

6. A stockbroker is a person who charges a commission or fee for buying or selling stocks.

7. A commission is a fee that a stockbroker charges for buying or selling stocks.

8. A prospectus explains how an investor's money will be spent.

9. When I contribute to an IRA, I can defer the taxes until later.

10. I want to buy stock that will make money.

Lesson 9: Interest

Pages 93–100

Pre-Lesson Vocabulary Practice

1. c
2. b
3. e
4. f
5. d
6. a
7. to measure
8. to calculate
9. annual
10. to earn
11. a deposit
12. the balance
13. a financial institution
14. $i = prt$
15. January–March
16. $\frac{1}{4}$
17. %
18. 0.04

Exercise 9

Part A

1. $i = prt = \$750 \times 0.025 \times 1 = \$18.75$

2. 8 months $= \frac{8}{12} = \frac{2}{3}$ year
   $i = prt = \$1,800 \times 0.025 \times \frac{2}{3} = \$30.00$

3. 3 years 6 months $= 3\frac{6}{12} = 3\frac{1}{2}$ or 3.5 years
   $i = prt = \$3,000 \times 0.085 \times 3.5 = \$892.50$
   new principal $= \$3,000 + \$892.50 = \$3,892.50$

4. 1 year 3 months $= 15$ months $= \frac{15}{12} = \frac{5}{4}$ or 1.25 years
   interest $= \$562.50 - \$500 = \$62.50$
   $r = \frac{i}{pt} = \frac{\$62.50}{\$500 \times 1.25} = \frac{62.5}{625} = \frac{1}{10} = 10\%$

5. $i = prt = \$2,000 \times 0.0525 \times 4 = \$420$

6. interest $= \$1,716 - \$1,600 = \$116$

   $t = \dfrac{i}{pr} = \dfrac{\$116}{\$1,600 \times 0.145} = \dfrac{116}{232} = 0.5$ year $=$
   6 months

7. 4 months $= \dfrac{4}{12} = \dfrac{1}{3}$ year

   $p = \dfrac{i}{rt} = \dfrac{\$195}{\$0.0975 \times \frac{1}{3}} = \dfrac{195}{0.0325} = \$6,000$

8. one and one half years $= 1.5$ years

   $r = \dfrac{i}{pt} = \dfrac{\$101.25}{\$900 \times 1.5} = \dfrac{101.25}{1,350} = 0.075 = 7.5\%$

9. nine months $= \dfrac{9}{12} = \dfrac{3}{4}$ or 0.75 year

   $i = prt = \$3,600 \times 0.0475 \times 0.75 = \$128.25$

10. interest $= \$470 - \$400 = \$70$

    $t = \dfrac{i}{pr} = \dfrac{\$70}{\$400 \times 0.07} = \dfrac{70}{28} = 2.5$ years or
    2 years 6 months

Part B

11. The interest for the first year is:
    $i = \$5,000 \times 0.09 \times 1 = \$450.$
    The principal at the end of the first year is:
    $\$5,000 + \$450 = \$5,450.$
    The interest for the second year is:
    $i = \$5,450 \times 0.09 \times 1 = \$490.50.$
    The principal at the end of the second year is:
    $\$5,450 + \$490.50 = \$5,940.50.$
    The interest for the third year is:
    $i = \$5,940.50 \times 0.09 \times 1 = \$534.645$ or
    $\qquad \$534.65.$
    The principal at the end of the third year is:
    $\$5,940.50 + \$534.65 = \$6,475.15.$

12. The interest for the first month is:
    $i = \$1,400 \times 0.035 \times \frac{1}{12} = \$4.083...$ or $\$4.08.$
    The principal at the end of the first month is:
    $\$1,400 + \$4.08 = \$1,404.08.$
    The interest for the second month is:
    $i = \$1,404.08 \times 0.035 \times \frac{1}{12} = \$4.095...$ or
    $\qquad \$4.10.$
    The principal at the end of the second month is:
    $\$1,404.08 + \$4.10 = \$1,408.18.$
    The interest for the third month is:
    $i = \$1,408.18 \times 0.035 \times \frac{1}{12} = \$4.107...$ or
    $\qquad \$4.11.$
    The principal at the end of the third month is:
    $\$1,408.18 + \$4.11 = \$1,412.29.$

Post-Lesson Vocabulary Reinforcement

1. Interest
2. Principal
3. Simple interest
4. Quarterly
5. Compound interest
6. Banks

7. $\underset{\underset{b}{\uparrow}}{\$750}$ at $\underset{\underset{c}{\uparrow}}{4\% \text{ annual interest}}$ $\underset{\underset{d}{\uparrow}}{\text{for 3 years}}$ $= \underset{\underset{a}{\uparrow}}{\$90}$

8. T

9. F A bank pays a customer interest for using the customer's money.

10. T

11. F Rate is the percent used to calculate the interest.

12. F Time is usually a number of years or a fraction of a year.

13. T

Language Builder

Activity A

1. of
2. on
3. in

Activity B

4. $\checkmark$ Interest is paid on the principal.

5. ___ The principal of $1,000 has earned 6% annual interest for 9 months.

6. $\checkmark$ Compound interest is calculated at regular intervals.

7. ___ Not surprisingly, compound interest is complicated.

8. $\checkmark$ The percent used to calculate the interest is called "rate."

Lesson 10: Loans

Pages 101–110

Exercise 10

Part A

1. 9 months $= \dfrac{9}{12} = \dfrac{3}{4}$ or 0.75 year
   $i = prt = \$500 \times 0.1 \times 0.75 = \$37.50$

2. $\$500 + \$37.50 = \$537.50$

3. 24 months $= 2$ years
   interest $= \$3,690 - \$3,000 = \$690$
   $r = \dfrac{i}{pt} = \dfrac{\$690}{\$3,000 \times 2} = \dfrac{690}{6,000} = 0.115 = 11.5\%$

Part B

4. $i = prt = \$6{,}500 \times 0.085 \times 4 = \$2{,}210$

5. total $= \$6{,}500 + \$2{,}210 = \$8{,}710$

6. 4 years $=$ 48 months
   $\$8{,}710 \div 48 = \$181.458...$ or $\$181.46$

Part C

7. 60 days $= \frac{60}{360} = \frac{1}{6}$ year
   $i = prt = \$800 \times 0.09 \times \frac{1}{6} = \$12$
   proceeds $=$ principal $-$ interest $=$
   $\$800 - \$12 = \$788$

8. interest $=$ principal $-$ proceeds $=$
   $\$2{,}000 - \$1{,}975 = \$25$
   $t = \frac{i}{pr} = \frac{\$25}{\$2{,}000 \times 0.15} = \frac{25}{300} = \frac{1}{12}$ year
   or 1 month or 30 days

9. interest $=$ principal $-$ proceeds $=$
   $\$900 - \$878.25 = \$21.75$
   time $= \frac{60}{360} = \frac{1}{6}$ year
   $r = \frac{i}{pr} = \frac{\$21.75}{\$900 \times \frac{1}{6}} = \frac{21.75}{150} = 0.145 = 14.5\%$

10. $i = prt = \$3{,}000 \times 0.185 \times 2 = \$1{,}110$
    total $= \$3{,}000 + \$1{,}110 = \$4{,}110$
    2 years $=$ 24 months
    $\$4{,}110 \div 24 = \$171.25$

Part D

11. 7.5% for a 48-month loan for a new car

12. 9.5% for a 36-month loan for a used car 6 years old or older

13. 8.75% for a 60-month loan for a used car less than 6 years old

14. No. A 48-month loan for used cars 6 years old or older is N/A, not available.

15. Interest rate $=$ 7.75% for a 60-month loan for a new car
    $i = prt = \$16{,}000 \times 0.0775 \times 5 = \$6{,}200$

16. Interest rate $=$ 8.25% for a 36-month loan for a used car less than 6 years old
    $i = prt = \$4{,}500 \times 0.0825 \times 3 = \$1{,}113.75$
    total $= \$4{,}500 + \$1{,}113.75 = \$5{,}613.75$

17. Interest rate $=$ 7.25% for a 36-month loan for a new car
    $i = prt = \$18{,}000 \times 0.0725 \times 3 = \$3{,}915$
    total $= \$18{,}000 + \$3{,}915 = \$21{,}915$
    monthly payment $= \$21{,}915 \div 36 = \$608.75$

Part E

18. $\$2{,}400 \div 6 = \$400$

19. $i = prt = \$2{,}400 \times 0.135 \times \frac{1}{12} = \$27$
    payment $= \$400 + \$27 = \$427$

20. new principal $= \$2{,}400 - \$400 = \$2{,}000$
    $i = prt = \$2{,}000 \times 0.135 \times \frac{1}{12} = \$22.50$
    payment $= \$400 + \$22.50 = \$422.50$

Post-Lesson Vocabulary Reinforcement

1. c
2. a
3. h
4. e
5. g
6. f
7. d
8. b
9. major purchase
10. mortgage loan
11. Interest rates
12. car loan
13. collateral
14. term of the loan

Language Builder

Activity A

1. The total that <u>Sandra</u> <u>will</u> <u>have</u> to pay is $6,200.

2. Find the proceeds that <u>Ramona</u> <u>will</u> <u>receive</u> from the bank.

3. What total amount of interest <u>will</u> <u>Carlos</u> <u>pay</u> on his loan?

4. What are the proceeds that <u>Ernesto</u> <u>will</u> <u>receive</u> on the loan?

5. In total, how much interest <u>will</u> <u>Helen</u> <u>pay</u> on her loan?

Activity B

6. Find the total amount, including interest, that Carlos will have to pay.

7. What total amount, including interest, will Juan owe the lender?

8. How much interest will Anna and Tom pay on their loan?

9. You will need formulas to solve problems in this lesson.

Activity C

10. What is the interest rate for a four-year loan on a new vehicle?

11. How much will Iwai pay for the car in all?

12. What rate of interest did Silvia pay on her loan?

13. How much interest did Richard owe his sister?

14. What is the interest rate on a three-year loan for a used car?

Lesson 11: Installment Buying

Pages 111–118

Exercise 11

Part A

1. total = $80 + (12 × $30) = $80 + $360 = $440

2. interest = $440 – $399 = $41

   rate $= \frac{i}{pt} = \frac{\$41}{\$360 \times 1} = 0.1138...$ or 11.4%

3. monthly payment = $468 ÷ 6 = $78

4. total = $100 + $468 = $568
   interest = $568 – $499.99 = $68.01

   rate $= \frac{i}{pt} = \frac{\$68.01}{\$468 \times 0.5} = \frac{\$68.01}{234} = 0.2906...$
   or 29.1%

5. total = $75 + (9 × $65) = $75 + $585 = $660

6. interest = $660 – $598 = $62

   rate $= \frac{i}{pt} = \frac{\$62}{\$585 \times 0.75} = \frac{62}{438.75} = 0.1413...$
   or 14.1%

7. total = $100 + (18 × $36) = $100 + $648 = $748

8. interest = $748 – $679 = $69

   rate $= \frac{i}{pt} = \frac{\$69}{\$648 \times 1.5} = \frac{69}{972} = 0.0709...$ or 7.1%

Part B

9. monthly payment = $711 ÷ 9 = $79

10. total = $100 + $711 = $811
    interest = $811 – $749.99 = $61.01

    rate $= \frac{i}{pt} = \frac{\$61.01}{\$711 \times 0.75} = \frac{61.01}{533.25} = 0.1144...$ or
    11.4%

11. total = $125 + (12 × $66) = $125 + $792 = $917

12. interest = $917 – $820 = $97

    rate $= \frac{i}{pt} = \frac{\$97}{\$792 \times 1} = 0.1224...$ or 12.2%

13. total = $200 + (24 × $59) = $200 + $1,416 = $1,616

14. interest = $1,616 – $1,399 = $217

    rate $= \frac{i}{pt} = \frac{\$217}{\$1,416 \times 2} = \frac{217}{2,832} = 0.0766...$ or 7.7%

Part C

15. total = $3,000 + (72 × $255) = $3,000 + $18,360 = $21,360

16. interest = $21,360 – $16,750 = $4,610

    rate $= \frac{i}{pt} = \frac{\$4,610}{\$18,360 \times 6} = \frac{4,610}{110,160} = 0.0418...$ or 4.2%

17. $\frac{\text{total price}}{\text{list price}} = \frac{\$21,360}{\$16,750} = 1.2752...$ or 127.5%

Post-Lesson Vocabulary Reinforcement

1. durable goods
2. products
3. examples
4. paying for
5. installment plan
6. agreement
7. down payment
8. balance due
9. collects
10. owns
11. used pickup truck
12. personal computer
13. 12-speed mountain bike
14. high definition television
15. digital camera
16. side-by-side refrigerator

Language Builder

Activity A

1. between
2. between
3. on
4. over

Activity B

5. B
6. C
7. B
8. A

Activity C

Answers will vary.

Lesson 12: Credit

Pages 119–128

Pre-Lesson Vocabulary Practice

1. a   2. c   3. b   4. d

Definitions are provided below. Sentences will vary.

5. retailers/retail stores – sellers and places where merchandise is sold

6. widely – commonly, usually, in many different ways

7. money order – a paper that one can purchase for a specific amount of money that can be used as that amount

8. finance charges – costs for using money that has not yet been paid back

9. credit card companies – companies that issue credit cards

10. cash advances – cash one can borrow and receive at an ATM by using one's credit card

11. to fall behind – to be late with

12. restrict – limit

13. theft – the taking of something that does not belong to one

14. identity – a person's name and other personal information

15. reputable – dependable, reliable

Exercise 12

Part A

1. total purchases = $32.06 + $28.79 + $56.28 + $43.01 = $160.14

2. unpaid balance minus payment = $795.23 – $250 = $545.23
   finance charge = $545.23 × 0.119 ÷ 12 = $5.41

3. new balance = $795.23 – $250 + $160.14 + $5.41 = $710.78

4. minimum payment = 0.15 × $710.78 = $106.62

Part B

5. $89.97 + $320.19 + $18.49 + $72.98 = $501.63

6. $1,073.88 – $300 = $773.88
   $773.88 × 0.165 ÷ 12 = $10.64

7. $1,073.88 – $300 + $501.63 + $10.64 = $1,286.15

8. 0.065 × $1,286.15 = $83.60

Part C

9. $216 + $14.25 + $39.56 + $45 + $89.91 = $404.72

10. $246.81 – $40 = $206.81
    $206.81 × 0.18 ÷ 12 = $3.10

11. $246.81 – $40 + $404.72 + $3.10 = $614.63

12. 0.025 × $614.63 = $15.37
    Since $20 is greater than $15.37, the minimum payment is $20.

Part D

13. 0.025 × $2,000 = $50

14. Month 4 $1,908.19 × 0.015 = $28.62, finance charge
    $$+\ \ 28.62$$
    $$\overline{\$1,936.81}$$
    $$-\ \ 50.00$$
    Month 5 $1,886.81 × 0.015 = $28.30, finance charge
    $$+\ \ 28.30$$
    $$\overline{\$1,915.11}$$
    $$-\ \ 50.00$$
    Month 6 $1,865.11 × 0.015 = $27.98, finance charge
    $$+\ \ 27.98$$
    $$\overline{\$1,893.09}$$
    $$-\ \ 50.00$$
    Month 7 $1,843.09 × 0.015 = $27.65, finance charge
    $$+\ \ 27.65$$
    $$\overline{\$1,870.74}$$
    $$-\ \ 50.00$$
    Month 8 $1,820.74 × 0.015 = $27.31, finance charge
    $$+\ \ 27.31$$
    $$\overline{\$1,848.05}$$
    $$-\ \ 50.00$$
    Month 9 $1,798.05 × 0.015 = $26.97, finance charge
    $$+\ \ 26.97$$
    $$\overline{\$1,825.02}$$
    $$-\ \ 50.00$$
    Month 10 $1,775.02 × 0.015 = $26.63, finance charge
    $$+\ \ 26.63$$
    $$\overline{\$1,801.65}$$
    $$-\ \ 50.00$$
    Month 11 $1,751.65 × 0.015 = $26.27, finance charge
    $$+\ \ 26.27$$
    $$\overline{\$1,777.92}$$
    $$-\ \ 50.00$$
    Month 12 $1,727.92 × 0.015 = $25.92, finance charge
    $$+\ \ 25.92$$
    $$\overline{\$1,753.84}$$
    $$-\ \ 50.00$$
    $1,703.84 = unpaid balance after 12 payments.

Although Joel paid 12 × $50 = $600 toward his original $2,000 charge, he still owes $1,703.84. In fact, he would owe even more if he made a late payment. Then the credit card company could charge a late fee and raise the interest rate on the finance charge.

Post-Lesson Vocabulary Reinforcement

1. Credit
2. Credit cards
3. Regular charge accounts
4. identity theft
5. Revolving charge accounts
6. d
7. e
8. b
9. a
10. c
11. F   Bank cards are sometimes called debit cards.
12. T
13. F   The monthly statement from a credit card company lists the activity in the account after the last statement.
14. T
15. F   The finance charges on unpaid balances vary widely.

Language Builder

Activity A

1. The annual interest rate can be as low as 6%.
2. What will be the finance charge on Renata's unpaid balance?
3. That balance will be due next month.
4. A high interest rate can be charged on some loans.
5. At the end of the year, Joel will still owe $1,908.19.
6. Credit cards can be used instead of cash in many stores.

Activity B

7. Anne's credit card company requires a minimum payment that is either 7.5% of the new balance or $20, whichever is <u>larger</u>.
8. No underline.
9. Mrs. Olmstead's credit card company charges a rate of 18% annual interest on unpaid balances

and requires a minimum payment of $20 or 2.5% of the unpaid balance, whichever is <u>greater</u>.

10. No underline.

Activity C

Answers will vary.

Lesson 13: Mortgages

Pages 129–138

Pre-Lesson Vocabulary Practice

1 – 9. Answers will vary.

10. the principal
11. the rate
12. the time
13. the first month's interest payment

Exercise 13

Part A

1. $734
2. $603
3. The largest monthly payment in the 7% column that is under $1,000 is $998.
$998 is the monthly payment for a $150,000 mortgage.
4. $402 − $307 = $95
5. The largest monthly payment under $1,200 in the 6% column is $1,199. This is the monthly payment for a $200,000 mortgage.

Part B

6. $i = prt = \$98,000 \times 0.05625 \times \frac{1}{12} = \$459.38$
7. $860.44 − $459.38 = $401.06
8. $98,000 − $401.06 = $97,598.94
9. $i = prt = \$97,598.94 \times 0.05625 \times \frac{1}{12} = \$457.50$
10. $860.44 − $457.50 = $402.94

Part C

11. $577
12. $1,302
13. $2,480
14. The largest monthly payment in the 30-yr column that is less than $1,000 is $961. $961 is the monthly payment for a $125,000 mortgage.

15. Mia's monthly payment is $620.
$10 \times 12 = 120$ months
$120 \times \$620 = \$74,400$

Part D

16. $i = prt = \$122,000 \times 0.0675 \times \frac{1}{12} = \$686.25$

17. $\$791.30 - \$686.25 = \$105.05$

18. $\$122,000 - \$105.05 = \$121,894.95$

Post-Lesson Vocabulary Reinforcement

1. b

2. a

3. c

4. g

5. d

6. e

7. f

8. T

9. F The amount that an adjustable rate mortgage can change each year is usually limited to one or two percentage points.

10. T

11. F The *cap* is another word for the maximum or highest amount

12. mortgage

13. cap

14. changes

15. year

Language Builder

Activity A

1. The buyer has <u>paid off</u> the mortgage.

2. The buyer <u>puts down</u> at least 20% of the asking price.

3. This amount <u>goes toward</u> interest on the new principal.

Activity B

4. paid off

5. put down

6. goes toward

7. pays off

Activity C

8. At a 7% annual interest rate what is the <u>largest</u> mortgage shown that Heather can borrow?

9. The <u>largest</u> amount that Kathy and Matt can afford is $1,200 a month.

10. What is the <u>least</u> amount Saori has to pay for her first month's interest payment?

Lesson 14: Utilities

Pages 139–153

Exercise 14

Part A

1. 07694 kwh

2. $07694 - 07605 = 89$ kwh
$89 \times \$0.13462 = \$11.981$ or $11.98

3. 24138 kwh

4. $24138 - 23989 = 149$ kwh
$149 \times \$0.113085 = \$16.849\ldots$ or $16.85

5. 60279 kwh

6. $60279 - 60096 = 183$ kwh
$183 \times \$0.167612 = \$30.672\ldots$ or $30.67

Part B

7. 1 month or 30 days

8. 2 months or 62 days

9. 137 kwh

10. 528 kwh

11. $12.30

12. $\$27.64 \div 2 = \$13.82$

13. $35231 - 35078 = 153$
$153 \times \$0.107076 = \$16.3826\ldots$ or $16.38
$\$12.30 + \$16.38 = \$28.68$

Part C

14. $4168 - 4098 = 70$ ccf

15. $70 \times \$0.266 = \$18.62$

16. $70 \times \$0.132 = \$9.24$

17. $\$7.00 + \$18.62 + \$9.24 = \$34.86$
$0.04 \times \$34.86 = \$1.3944\ldots$ or $1.39
$\$34.86 + \$1.39 = \$36.25$

Part D

18. $931 - 917 = 14$ hcf

19. $14 \times \$1.395 = \$19.53$
$\$11.75 + \$19.53 + \$10.25 = \$41.53$

Part E

20. Subtotal for basic service and calls:
$\$19.95 + \$6.44 + \$0.23 + \$0.59 = \$27.21$
Subtotal for custom calling features:
$\$3.00 + \$6.00 = \$9.00$
Subtotal for taxes and surcharges:
$\$0.86 + \$1.00 + \$1.26 + \$1.23 = \$4.35$
Total $= \$27.21 + \$9.00 + \$4.35 = \$40.56$

21. The $3.00 charge for call blocking and the $6.00 charge for voice mail are charges that should appear *only* if a customer requests them.

Post-Lesson Vocabulary Reinforcement

1. the first five places

2. abbreviation

3. Federal Communications Commission

4. utility

5. FCC Line Charge

6. therm

7. the symbol for 100

8. Cramming

9. Local Number Portability or LNP

10. 911 service fee

11. scams

12. Telecommunications relay

13. Slamming

14. public utility

15. kilowatt-hour

16. Usage

17. telephone

18. water and natural gas

Language Builder

Activity A

1. that provides a commodity

2. that measures the number of kilowatt-hours

3. that allows customers to keep their telephone number

4. When the hand on a dial is between two digits

Activity B

5–9. Answers will vary.

Activity C

10. d. should

11. a. will

12. b., c. can, cannot

Lesson 15: Renting or Buying a Home

Pages 153–162

Pre-Lesson Vocabulary Practice

1. head of household

2. U.S. Census Bureau

3. sign a lease

4. trailers/mobile homes

5. houseboats

6. Hotel/motel rooms

7. rental properties

8. landlord

9. tenant

10. financial responsibility

11. appreciate

12. depreciate

13. Rent

14. security deposit

15. agents

16. expenditure

17. lawyers

Exercise 15

Part A

1. $\$490 + \$490 + \$300 = \$1,280$

2. $\$490 + \$65 = \$555$
$12 \times \$555 = \$6,660$

3. $810 + $810 + $750 = $2,370
   $2,370 ÷ 3 = $790

4. $810 ÷ 3 = $270

5. 1.045 × $604.70 = $631.911... or $631.91

6. 0.3 × $31,400 = $9,420
   $9,420 ÷ 12 = $785

Part B

7. $28,500 + $6,500 = $35,000

8. 0.24 × $129,000 = $30,960

9. $129,000 – $30,960 = $98,040

10. $30,960 + $2,350 = $33,310

11. 0.3 × $35,000 = $10,500
    $10,500 ÷ 12 = $875
    Since their payment is less than $875, they have enough income.

Part C

12. Los Angeles, California

13. Dayton, Ohio

14. 0.2 × $224,700 = $44,940

15. $402,100 – $197,900 = $204,200

16. $\frac{$108,000}{$136,500}$ = 0.791... or about 79%

Part D

17. $14,500.00 + $26,738.99 = $41,238.99

18. $\frac{$41,238.99}{$145,000}$ = 0.284... or about 28%

19. 0.06 × $145,000 = $8,700

20. 12 × $691.93 = $8,303.16
    $\frac{$8,303.16}{$32,300}$ = 0.257... or about 26%

Post-Lesson Vocabulary Reinforcement

1. a	6. c
2. f	7. e
3. j	8. g
4. d	9. h
5. b	10. i

Answers will vary for problems 11–17, but should be similar to these:

11. to rent or to buy

12. the owner or landlord

13. to pay the monthly rent

14. because they are worth more and increase in value over time

15. a car

16. when a seller and a buyer agree on a price and the buyer has arranged for financing

17. in an escrow account

Language Builder

Activity A

1. No underlining

2. The head of a household has two choices when <u>deciding</u> how <u>to pay</u> for housing: <u>to rent</u> or <u>to buy.</u>

3. ...are <u>paying</u> off their mortgages.

4. After <u>paying</u> the initial charges, the renter's only financial responsibility is <u>to pay</u> the monthly rent.

5. <u>Buying</u> a home requires far more cash up front than <u>renting</u>.

Activity B

6. To rent or to buy a home is an important decision.

7. LEAVE BLANK

8. Renting a house doesn't require as much cash up front as buying one does.

9. To buy a home requires far more cash up front than to rent one.

10. LEAVE BLANK

Activity C

11. to use / using

12. Buying / Renting

13. doing / finishing / completing ... to rest / to watch TV / etc.

14. to buy / to rent ... to pay

Lesson 16: Car Expenses

Pages 163–172

Exercise 16

Part A

1. 920 miles ÷ 33 gallons = 27.87... or 27.9 mpg

Part B

2. $1,930 ÷ 723 gallons = $2.669... or $2.67

3. 12,164 miles ÷ 723 gallons = 16.82...or 16.8 mpg

4. 12,164 ÷ 3,000 = 4.05... or 4 oil changes
4 × $22.95 = $91.80

Part C

5. $726 ÷ 266 gallons = $2.729... or $2.73

6. 9,708 miles ÷ 266 gallons = 36.49... or 36.5 mpg

7. 9,708 miles ÷ 3 = 3,236 miles

Part D

8. 0.8 × 10,973 miles = 8,778.4 or 8,778 miles

9. 0.2 × 10,973 miles = 2,194.6 or 2,195 miles
or 10,973 − 8,778 = 2,195 miles

10. 8,778 miles ÷ 34 mpg = 258.1... or 258 gallons

11. 2,195 miles ÷ 26 = 84.4... or 84 gallons

12. total gallons = 258 + 84 = 342 gallons
342 × $2.899 = $991.458 or $991.46

Part E

13. $1,050 + $900 + $775 + $650 + $550 = $3,925

14. $5,999 − $3,925 = $2,074

Part F

15. $780 + $1,339 + $1,190 + $1,204 + $724 = $5,237

16. $8,800 ÷ $18,699 = 0.4706... or 47%

17. $5,237 + $8,800 = $14,037

Part G

18. depreciation = $6,298 − $5,173 = $1,125
$\frac{\$1,125}{\$6,298}$ = 0.178... or about 18%

19. depreciation = $30,699 − $18,949 = $11,750
$\frac{\$11,750}{\$30,699}$ = 0.382... or about 38%

Post-Lesson Vocabulary Reinforcement

1. more than

2. public transportation system

3. A car

4. fuel

5. division

6. fuel-efficient

7. expenses

8. depreciate

9. million

10. stores

11. frequently

12. worn

13. <u>o</u> <u>n</u> <u>e</u> <u>r</u> <u>w</u>

14. owner

Language Builder

Activity A

1. bought   3. pay

2. driven   4. spent

Answers will vary.

Activity B

5. has to   7. have to

6. should   8. must

Lesson 17: Insurance

Pages 173–186

Exercise 17

Part A

1. 12 × $16.75 = $201

2. $1,200 − $250 = $950

3. 1.04 × $165 = $171.60

Part B

4. 0.0072 × $133,500 = $961.20

5. 100% − 6% = 94%
0.94 × $961.20 = $903.528 or $903.53

6. 1.09 × $847 = $923.23

7. $784 − $721 = $63
$\frac{\$63}{\$721}$ = 0.087... or 8.7%

8. $6,649 − $4,947 = $1,702
$\frac{\$1,702}{\$4,947}$ = 0.344... or about 34%

9. $2,066 − $1,480 = $586
$\frac{\$586}{\$1,480}$ = 0.395... or about 40%

Part C

10. $224 + $4 + $110 + $169 + $96 + $102 + $6 = $711

11. $430 – $200 = $230

12. $\frac{\$224}{\$711} = 0.315...$ or about 32%

Part D

13. $(2 \times \$15) + \$240 = \$30 + \$240 = \$270$
There is only one $240 co-payment for the hospital stay.

14. A $15 co-payment covers both the checkup and the X-ray.
There is no co-payment for the blood test (laboratory service).
$15 + $15 = $30

15. $1.085 \times \$2,976.80 = \$3,229.828$ or $3,229.83

Part E

16. $12 \times \$63.00 = \$756.00$

17. $12 \times \$39.60 = \$475.20$

18. $20 \times 12 \times \$26.55 = \$6,372.00$

19. $12 \times \$97.20 = \$1,166.40$
$12 \times \$72.45 = \$869.40$
$\$1,166.40 – \$869.40 = \$297.00$

Part F

20. $6 \times \$42,000 = \$252,000$

Post-Lesson Vocabulary Reinforcement

1. b
2. c
3. f
4. d
5. a
6. g
7. e
8. h
9. j
10. i
11. deductible
12. Liability
13. first-party coverage

14. Comprehensive insurance
15. uninsured motorist
16. No-fault insurance
17. rider
18. Medicare
19. Medicaid

Language Builder

Activity A

1. An insurance policy is an agreement between a consumer <u>who</u> pays for the policy and the company <u>that</u> offers the insurance.

2. An insurance policy is a detailed document <u>that describes the coverage</u> <u>that</u> the insurer will pay in case of a financial loss.

3. <u>When</u> a policyholder makes a claim, he often has to pay a deductible.

Activity B

Possible answers:

4. ...protects against the high costs of treating illnesses and injuries...

5. ...is an insurance policyholder...

6. ...pay a certain amount of money.

Activity C

7. There <u>is</u> <u>usually</u> a deductible for collision insurance.

8. The value of a car <u>depreciates</u> <u>quickly</u>.

9. Term life insurance may be <u>renewed</u> <u>annually</u>.

Part D

Answers will vary.

Lesson 18: Taxes

Pages 187–196

Exercise 18

Part A

1. $0.075 \times \$431 = \$32.325$ or $32.33
$\$32.33 + \$3.00 = \$35.33$

2. $0.075 \times \$532 = \$39.90$
one trip = $\$532.00 + \$39.90 + \$3.00 + \$3.00 = \$577.90$
two trips = $2 \times \$577.90 = \$1,155.80$

3. $0.075 \times \$236.50 = \$17.7375$ or $\$17.74$
   The whole trip has 4 segments for an additional $4 \times \$3 = \$12$.
   Total $= \$236.50 + \$17.74 + \$12.00 = \$266.24$

Part B

4. Georgia

5. Rhode Island

6. $\$0.183 + \$0.25 = \$0.433$

7. $\frac{15\cent}{30\cent} = \frac{1}{2}$

8. $\$0.183 + \$0.235 = \$0.418$
   $\frac{\$0.418}{\$2.55} = 0.163...$ or about 16%

Part C

9. $110 - 90 = 20$ pounds
   $20 \times \$0.50 = \$10$
   $\$10.50 + \$10.00 = \$20.50$
   $4 \times \$20.50 = \$82.00$

10. $120 - 90 = 30$ pounds
    $30 \times \$0.50 = \$15$
    $\$10.50 + \$15.00 = \$25.50$
    $50 \times \$25.50 = \$1,275.00$

11. $0.0218 \times \$127,900 = \$2,788.22$

12. $\frac{\$1,812}{\$149,000} = 0.012...$ or 1.2%

13. $0.0295 \times \$12,500 = \$368.75$

Part D

14. $\$78,200 \div \$1000 = 78.2$
    $\$4.523 \times 78.2 = \$353.698...$ or $\$353.70$

15. $\$3.738 \times 78.2 = \$292.311...$ or $\$292.31$

16. $\$7.916 \times 78.2 = \$619.031...$ or $\$619.03$

17. $1.048 \times \$619.03 = \$648.743...$ or $\$648.74$

Part E

18. $\$0.00657 \times 115,300 = \$757.521...$ or $\$757.52$

19. $\$0.01493 \times 115,300 = \$1,721.429...$ or $\$1,721.43$

20. $0.59 \times \$26,300,000 = \$15,517,000$
    $\$15,517,000 \div 12,000 = \$1,293...$ or about $\$1,300$

Post-Lesson Vocabulary Reinforcement

1. revenue
2. excise tax
3. mills
4. Property taxes
5. assessor
6. property tax

7. market value

8. mill

9. real property or real estate

10. millage

11. T

12. F Taxes are always imposed on airfares.

13. T

14. T

15. F Property taxes are imposed on buildings and pieces of land.

16. T

Language Builder

Activity A

1. is called, <u>income</u>

2. are taxed, <u>Land and buildings</u>

3. is assessed, <u>Ms. Kay's property</u>

4. be based, <u>Property tax</u>

5. are measured, <u>property tax rates</u>

6. was given, <u>tax rate</u>

Activity B

7. Mr. and Mrs. Gonzalez **bought** round-trip tickets from Charlotte, North Carolina, to Minneapolis, Minnesota.

8. One round-trip fare **was** $532.

9. The return flight **had** a stopover in Chicago.

10. Bob **paid** $2.55 for a gallon of gasoline in Maryland.

Activity C

11. The dollar sign ($) **disappears** in that stated assessed value of the Martin's house.

12. The property tax rate **is** given as a number of dollars per $1,000 of value.

13. This lesson **includes** examples of excise taxes and property taxes.

14. There **are** two segments to Aaron's flight.

15. How much **does** the owner get for the property?

16. Nina **pays** a lot for travel.

Lesson 19: Income Tax

Pages 197–210

Exercise 19

Part A

1. 35%
2. 25%
3. 15%
4. $174,700
5. $28,400

Part B

6. Use Schedule Y-1
$63,490 – $56,800 = $6,690
$6,690 × 0.25 = $1,672.50
$1,672.50 + $7,820 = $9,492.50

7. Use Schedule Z
$29,215 – $10,000 = $19,215
$19,215 × 0.15 = $2,882.25
$2,882.25 + $1,000 = $3,882.25

8. Use Schedule Y-2
$43,760 – $28,400 = $15,360
$15,360 × 0.25 = $3,840
$3,840 + $3,910 = $7,750

9. Use Schedule X
$72,500 – $68,800 = $3,700
$3,700 × 0.28 = $1,036
$1,036 + $14,010 = $15,046

10. Use Schedule Y-1
$62,850 – $56,800 = $6,050
$6,050 × 0.25 = $1,512.50
$1,512.50 + $7,820 = $9,332.50

Part C

11. $5,004
12. $5,004 – $4,500 = $504 owed to IRS
13. $4,484
14. $5,200 – $4,484 = $716 refund
15. $4,201
16. $4,201 – $4,150 = $51 owed to IRS
17. $5,054
18. $5,054 – $4,980 = $74 owed to IRS
19. $5,141
20. $5,700 – $5,141 = $559 refund

Part D

21.

Post-Lesson Vocabulary Reinforcement

1.	b	**9.**	IRS
2.	c	**10.**	extension
3.	a	**11.**	tax forms
4.	e	**12.**	itemized
5.	h	**13.**	charitable gifts
6.	g	**14.**	tax due
7.	f	**15.**	refund
8.	d	**16.**	form 1040X

Possible answers:

17. From the gross income, subtract contributions to any retirement accounts, alimony payments, and interest payments on student loans.

18. Employers

19. April 15th

20. Medical expenses OR dental expenses / state taxes / local taxes / real estate taxes / mortgage interest / charitable gifts / job-related expenses

Language Builder

Activity A

1.	and	**4.**	and
2.	or	**5.**	or
3.	but	**6.**	but

Activity B

7. that lists the income and the tax that is due

8. a taxpayer made in a year

9. that was already paid during the year

10. If Martin is single

11. who can help taxpayers to complete their tax forms

Lesson 20: Writing a Budget

Pages 212–221

Part A

1. $1,100 \div $40 per month = 27.5 months or 2 years 3.5 months

2. $1\frac{1}{2}$ years = 18 months
$750 \div 18 = $41.67 per month

3. $560 \div $35 per month = 16 months

4. 30 − 28 = 2 years = 24 months
$1,400 \div 24 = $58.33 per month

Part B

5. Nevada

6. North Dakota

7. 5 × $362 = $1,810 for meals and lodging
5 × $30 = $150 for incidentals
total = $1,810 + $150 + $80 = $2,040
$2,040 \div 7 months = $291.43 per month

8. 14 × $417 = $5,838
14 × $279 = $3,906
$5,838 − $3,906 = $1,932
or $417 − $279 = $138 per day
14 × $138 = $1,932

9. meals & lodging = 7 × $252 = $1,764
airfare = 4 × $200 = $800
incidentals = 7 × $30 = $210
parking = 7 × $10 = $70
total = $1,764 + $800 + $210 + $70 + $20 +
$20 + $40 = $2,924
$2,924 \div 12 months = $243.67 per month

Part C

10. total tuition = 4 × $4,350 = $17,400
5 years = 5 × 12 months = 60 months
$17,400 \div 60 = $290 per month

Part D

11. monthly housing expenses:
mortgage = $639
heating oil = $816 \div 12 = $68
electricity = $49
telephone = $64
insurance = $689 \div 12 = $57.42 or about $57.
real estate taxes = $1,628 \div 12 = $135.67 or
about $136.
total monthly housing expenses = $1,013.

12. total monthly income = $36,780 \div 12 = $3,065

13. monthly housing expenses \div monthly income
= $1,013 \div $3,065 = 0.3305... or 33.1%.

14. loan payments = 12 × $431 = $5,172
total transportation costs =
$5,172 + $847 + $528 + $80 = $6,627
transportation costs \div take-home income =
$6,627 \div $36,780 = 18%

Part E

15.
Housing	$9,605
Transportation	$4,940
Food	$6,365
Clothes	$2,483
Health care	$1,691
Entertainment	$1,988
Insurance & pensions	$2,073
Other	$3,316
Total	$32,461

16. gross income = $32,461 + $6,230 = $38,691

17.
Housing	$9,605 ÷ $32,461 = 29.6%
Transportation	$4,940 ÷ $32,461 = 15.2%
Food	$6,365 ÷ $32,461 = 19.6%
Clothes	$2,483 ÷ $32,461 = 7.6%
Health care	$1,691 ÷ $32,461 = 5.2%
Entertainment	$1,988 ÷ $32,461 = 6.1%
Insurance & pensions	$2,073 ÷ $32,461 = 6.4%
Other	$3,316 ÷ $32,461 = 10.2%

18. housing + transportation + food = 29.6% + 15.2% + 19.6% = 64.4%

19. Rhonda and Malcolm spent more on food, clothes, and entertainment than the average. They spent less on housing, transportation, health care, insurance & pensions, and other.

Posttest 1

1. 35 × $17.50 = $612.50

2. 0.0765 × $612.50 = $46.856... or $46.86

3. $824.62 + $0.48 – $2.00 = $823.10

4. $606.73 + $432.70 – $76.33 – $140.00 = $823.10

5. $34.99 + $28.99 = $63.98
   1.045 × $63.98 = $66.8591 or $66.86

6. 20% off means the price 100% – 20% = 80% of the original price
   0.8 × $169 = $135.20
   1.065 × $135.20 = $143.988 or $143.99

7. $7.22 ÷ 0.85 = $8.494... or $8.49 per pound

8. $2.99 ÷ 11.5 = $0.26 per ounce

9. $19.90 + $14.90 + $1.90 + $1.90 + $2.45 = $41.05
   1.06 × $41.05 = $43.513 or $43.51

10. $\frac{\$7}{\$41.05}$ = 0.1705... or about 17%

11. initial cost = $2.00
    7.5 miles × $1.50 per mile = $11.25
    2 × $2.00 = $4 for delay

total = $2.00 + $11.25 + $4.00 = $17.25
1.15 × $17.25 = $19.8375 or $19.84
The customer would probably give the driver $20.

12.
```
  3:40 P.M. = 15:40 = 14 hours 100 minutes
– 9:50 A.M. =          9 hours  50 minutes
                       5 hours  50 minutes
```

13. $P = 2l + 2w = 2(36) + 2(10) = 72 + 20 =$ 92 feet
    92 × $1.39 = $127.88

14. living room area is $A = lw = 20 \times 14 =$ 280 square feet
    dining room area is $A = lw = 14 \times 10 =$ 140 square feet
    total area = 280 + 140 = 420 square feet
    1 square yard = 3 × 3 = 9 square feet
    420 ÷ 9 = 46.666... or about 46.7 square yards
    46.7 × $25.90 = $1,209.53

15. 200 × $15.30 = $3,060
    200 × $19.25 = $3,850
    $3,850 – $3,060 = $790

16. 0.004 × $3,850 = $15.40
    $20.00 + $15.40 = $35.40

17. interest = $2,240 – $2,000 = $240
    $t = \frac{i}{pr} = \frac{\$240}{\$2,000 \times 0.08} = \frac{\$240}{\$160} = 1.5$ years or 18 months

18. $i = prt = \$1,500 \times 0.18 \times \frac{1}{12} = \$22.50$

19. $i = prt = \$7,500 \times 0.085 \times 2 = \$1,275$

20. $i = prt = \$5,000 \times 0.165 \times 3 = \$2,475$
    total = $5,000 + $2,475 = $7,475
    3 years = 3 × 12 = 36 months
    $7,475 ÷ 36 = $207.638... or $207.64 per month

21. $200 + (18 × $45) = $200 + $810 = $1,010

22. interest = $1,010 – $799 = $211
    $rate = \frac{i}{pt} = \frac{\$211}{\$810 \times 1.5} = \frac{\$211}{\$1,215.00} = 0.1736...$ or 17.4%

23. $2,736 × 0.195 × $\frac{1}{12}$ = $44.46

24. $2,736 × 0.025 = $68.40

25. $i = prt = \$85,000 \times 0.075 \times \frac{1}{12} = \$531.25$

26. $880.67 – $531.25 = $349.42

27. usage = 37709 – 37426 = 283 kwh
    283 × $0.148314 = $41.972... or $41.97

28. usage = 3287 − 3251 = 36 ccf
    36 × $0.312 = $11.232 or $11.23
    36 × $0.126 = $4.536 or $4.54
    subtotal = $11.23 + $4.54 + $6.00 = $21.77
    1.045 × $21.77 = $22.749... or $22.75

29. 0.3 × $37,500 = $11,250
    $11,250 ÷ 12 = $937.50 per month

30. increase in value = $129,000 − $88,000 = $41,000
    $\frac{\$41,000}{\$88,000} = 0.4659...$ or 46.6%

31. 74.650 − 62,412 = 12,238 miles
    12,238 miles ÷ 486 gallons = 25.18...
    or 25.2 mpg

32. decrease in value = $19,500 − $6,600 = $12,900
    $\frac{\$12,900}{\$19,500} = 0.6615$ or about 66%

33. 0.0068 × $142,900 = $971.72

34. 1.055 × $3,241 = $3,419.255 or $3,419.26

35. 2 segments = 2 × $3 = $6
    0.075 × $407 = $30.525 or $30.53
    $30.53 + $6.00 = $36.53

36. $99,500 ÷ $1,000 = 99.5
    99.5 × $6.853 = $681.873 or $681.87

37. $46,528 − $14,000 = $32,528
    0.15 × $32,528 = $4,879.20
    $4,879.20 + $1,400.00 = $6,279.20

38. 12 × $608 = $7,296
    $7,296.00 − $6,279.20 = $1,016.80 refund

39. $31,095 ÷ 12 = $2,591.25 monthly income
    $638 + $47.50 + $66.33 = $751.83
    $\frac{\$751.83}{\$2,591.25} = 0.2901...$ or 29.0%

40. $132 + $136 = $268
    10 × $268 = $2,680
    $2,680 ÷ 12 = $223.33 per month

Posttest 2: Write a Budget
A. HOUSING

mortgage payment = $665 per month

water and sewer = $80 per quarter × 4 = $320 per year

$320 ÷ 12 = $26.67 per month

gas and electricity = $43 per month

telephone = $78.56 + $49.23 + $82.05 = $209.84

$209.84 ÷ 3 = $69.95 average per month

trash pickup = $4 per week × 4 = $16 per month

property tax = $147,500 ÷ $1,000 = 147.5

$11.94 × 147.5 = $1,761.15

$1,761.15 ÷ 12 = $146.76 per month

homeowner's insurance = 0.0069 × $147,500 = $1,017.75

$1,017.75 ÷ 12 = $84.81 per month

garage roof = $1,050 ÷ 12 = $87.50 per month

HOUSING subtotal = $1,139.69 per month

B. TRANSPORTATION

interest on car loan = $5,500 × 0.085 × 4 = $1,870

total car loan = $5,500 + $1,870 = $7,370

$7,370 ÷ 48 months = $153.54 per month

car insurance = $1,149 + $726 = $1,875 per year

$1,875 ÷ 12 = $156.25 per month

servicing cars = 4 × $25 = $100 per year

$100 ÷ 12 = $8.33 per month

repairs on old car = $500 ÷ 12 = $41.67

gasoline = 808 × $2.55 = $2,060.40 per year

$2,060.40 ÷ 12 = $171.70 per month

airfare = $824 ÷ 12 = $68.67

TRANSPORTATION subtotal = $600.16 per month

C. FOOD

	groceries	restaurants
September	$338.28	$127.42
October	350.82	88.78
November	483.56	42.89

$1,172.66 + $259.09 = $1,431.75 total

$1,431.75 ÷ 3 = $477.25 average per month

FOOD subtotal = $477.25 per month

D. CLOTHING

$84.72 + $105.63 + $117.53 + $96.52 + $49.38 + $63.27 = $517.05

$517.05 ÷ 3 = $172.35 average per month

CLOTHING subtotal = $172.35 per month

E. HEALTHCARE

office co-payments = 8 × $15 = $120

$120 ÷ 12 = $10 per month

hospital co-payment = $240 ÷ 12 = $20 per month

blood pressure medication = $39.76 per month

miscellaneous medicine = $100 ÷ 12 = $8.33 per month

dentist = 8 × $45 = $360 per year

$360 ÷ 12 = $30 per month

glasses = $275 + $195 = $470

$470 ÷ 12 = $39.17 per month

HEALTHCARE subtotal = $147.26 per month

F. ENTERTAINMENT

cable TV = $21.95 per month

Internet access = $23.90 per month

weekly newsmagazine = $49.90 ÷ 12 = $4.16 per month

daily newspapers = 21 × $0.75 = $15.75 per month

Sunday newspapers = 4 × $3.50 = $14.00 per month

movies = $16.00 per month

new television = $599 ÷ 12 = $49.92

new bicycle = $249 ÷ 12 = $20.75

ENTERTAINMENT subtotal = $166.43 per month

G. PERSONAL INSURANCE & PENSIONS

Lena's life insurance premium = $39.60 per month

David's life insurance premium = $50.85 per month

pensions & Social Security = $3,151.80 ÷ 12 = $262.65 per month

PERSONAL INSURANCE & PENSIONS subtotal = $353.10 per month

H. OTHER

credit card interest & fees: $0 + $100 + $250 = $350 for three months

$350 ÷ 3 = $116.67 average per month

child care = $30 per week × 4 = $120 per month

gifts & donations = $400 + $900 = $1,300 per year

$1,300 ÷ 12 = $108.33 per month

new refrigerator = $699 ÷ 12 = $58.25 per month

OTHER subtotal = $403.25 per month

SUMMARY BUDGET

	Monthly Expense			Percent
A. HOUSING	$1,140	÷	$3,458 =	33.0%
B. TRANSPORTATION	600	÷	$3,458 =	17.4%
C. FOOD	477	÷	$3,458 =	13.8%
D. CLOTHING	172	÷	$3,458 =	4.9%
E. HEALTHCARE	147	÷	$3,458 =	4.3%
F. ENTERTAINMENT	166	÷	$3,458 =	4.8%
G. PERSONAL INSURANCE & PENSIONS	353	÷	$3,458 =	10.2%
H. OTHER	403	÷	$3,458 =	11.7%
Total	$3,458		=	100.0%

Compared to the average household expenditures in the table on page 213, the Ross family spends less on transportation, healthcare, and entertainment. They spend more on housing, food, clothing, personal insurance & pensions, and "other." Families with young children are likely to spend more on food and clothing. The higher expenses in the "other" category may point out a problem. David and Lena should monitor the amount that they leave unpaid on their monthly credit card bills. And they should save more each month for the education expenses that will come when Catherine and Alex get older.

GLOSSARY

A

adjustable rate An interest rate that changes, usually from year to year

approximate Another word for estimate. As an adjective: close to or about

area The amount of surface on a flat figure. The area of the floor of a room that is 15 feet long and 10 feet wide is $15 \times 10 = 150$ square feet.

assessed value The value of a property that is used for determining property tax

asset Anything of value, such as a house, a car, or stocks and bonds

average The sum of a set of values divided by the number of values in the set. The average of 11, 15, and 16 is the sum (42) divided by the number of numbers (3). $42 \div 3 = 14$. *Average* is another word for *mean*.

B

balance The difference between an amount that is owed and the amount that has already been paid

bond A way for companies and government to raise money; a bond holder receives yearly interest from the company or government agency that issues the bond.

budget A detailed summary of estimated expenses over a period of time

C

claim A request for payment by an insurance policyholder from the insurance company

collateral Something of value used to secure a loan

commission A fee for services, often a percentage of the total cost

compound interest A percent of money paid on both principal and any accumulated interest

co-payment A dollar amount that an insured person pays for medical services

coupon rate The percent that the issuer of a bond is required to pay a bond holder

credit A dollar amount that is added to an account; the opposite of a debit

D

debit A dollar amount that is subtracted from an account

decimal A part of a whole in which the whole is divided into tenths, hundredths, thousandths, and so on; for example, $0.09 = \frac{9}{100}$.

decimal point A dot that separates whole numbers from decimals. In 25.3 the decimal point separates the whole number 25 from the fraction three-tenths.

deductible A fee or an expense that can be subtracted; for example, certain expenses are deductible from income tax. Also the amount of a claim that an insurance policyholder must pay.

deduction An amount that is subtracted from a gross amount, such as union dues from a paycheck

demand loan or demand note A loan that is secured with collateral

denominator The bottom number or divisor in a fraction. In $\frac{2}{3}$ the denominator is 3. The denominator tells the number of parts in the whole.

dependent A person for whom a taxpayer has financial responsibility

depreciation A decrease in value over a period of time

difference The answer to a subtraction problem. For the problem $0.1 - 0.06$, the difference is 0.04.

digit One of the ten number symbols. The digits are 0, 1, 2, 3, 4, 5, 6, 7, 8, and 9.

dimension A measurement of distance such as the length of a rectangle

discount As a verb, to subtract from a price; as a noun, an amount that is subtracted from a price.

dividend The number in a division problem that is divided by another number. In the problem $1.5 \div 0.3$, the dividend is 1.5. In business, dividend is a share of the earnings of a company.

divisor The number in a division problem that divides into another number. In the problem $1.5 \div 0.3$, the divisor is 0.3.

down payment An initial payment toward a purchase, such as the amount first paid toward the purchase of a house

E

equity Another word for *stock*

estimate As a noun, an approximate value; as a verb, to find an approximate value

excise tax A tax imposed by a government for specific goods and services

exemption An amount of money that is free from taxation

F

finance charge A percent of an unpaid balance

fixed rate An interest rate that remains unchanged

formula A mathematical rule written with an $=$ sign. The formula for finding the cost of something is $c = nr$ where c is the cost, n is the number of items, and r is the rate or cost for one item.

fuel efficiency The amount of gasoline that a vehicle requires to travel a certain distance

G

gratuity A sum of money given in appreciation for a service

gross An entire amount, like a wage before any deductions

H

horizontal Parallel to the horizon; for a geometric figure drawn on a page, the horizontal dimension is the greatest left-to-right measurement.

household A person or group of people, related or not, who occupy a single dwelling

I

installment A regular payment such as a monthly loan payment

interest A percentage of an amount of money

K

kilowatt-hour A measurement of electrical usage; one kilowatt-hour is enough electricity to light a 100-watt bulb for 10 hours.

L

lease A contract that allows the use of a space, such as an apartment, for a specific length of time

liability A legal obligation; liability insurance is for medical payments and property damage to another party.

linear In a straight line, like the length of a rectangle

list price A published price before any discounts are subtracted

M

market value The price that an owner can expect to receive when selling a property

mean Another word for *average*; a mean is the sum of a set of numbers divided by the number of numbers in the set.

median A number that represents the middle value of a set of numbers

mill One-thousandth of a dollar; often used to measure a property tax rate

mortgage An installment loan, usually for the purchase of a house or land

mutual fund An investment company that invests the cash of many shareholders in a certain way.

N

net Remaining after deductions are made, such as a net salary

net asset value (NAV) The value of a mutual fund's holdings divided by the number of shares

numerator The top number in a fraction. In the fraction $\frac{4}{5}$, the numerator is 4.

O

overtime Working hours beyond the usual weekly number of hours

P

party A person involved in a legal or financial transaction

per annum By the year

percent One-hundredth or $\frac{1}{100}$. The symbol for percent is %. A whole amount is 100%. Half of a whole amount is 50%.

perimeter A measure of the distance around a flat figure. A room that is 15 feet long and 10 feet wide has a perimeter of 50 feet.

place value The number that a digit stands for. In 356, the digit 5 has a value of 50 because it is in the tens place.

policy A written insurance contract

premium The cost of an insurance policy

prime The interest rate that lenders charge their best customers; the prime rate is often used to measure the interest rate on mortgages.

principal The amount of money on which interest is calculated

proceeds The amount, minus interest, that a borrower receives on a short-term loan

product The answer to a multiplication problem. The product of 8 × 5 is 40.

promissory note A written promise to pay a certain amount of money in a certain amount of time

Q

quotient The answer to a division problem. In the problem 6.8 ÷ 2, the quotient is 3.4.

R

rate A measurement in terms of another measurement. Rates often contain the word *per*, such as miles per gallon or dollars per month.

real estate A building or a piece of land or both

rectangle A four-sided figure with four square corners. A page from a newspaper is a rectangle.

reducing Expressing a fraction with smaller numbers than the numbers in the original fraction. The fraction $\frac{9}{12}$ reduces to $\frac{3}{4}$. The reduced form of a fraction is equal in value to the original fraction.

rounding An estimate that is close to an original number. Rounded to the nearest hundred dollars, $2,694 is $2,700.

S

salary Payment for work, often on a yearly basis

share One unit of ownership of a company

simple interest A percent of money paid on only the principal

stock An ownership share in a company

subtotal A partial total; the sum of a set of numbers in a larger set

sum The answer to an addition problem. The sum of 1.6 and 3.8 is 5.4.

T

term A length of time, such as the number of years of a mortgage

tip As a noun, a sum of money beyond the total of a bill that is given in appreciation for a service; as a verb, to give a sum of money in appreciation for a service

total Another word for *sum,* the answer to an addition problem

U

unit of measurement A label for measuring some aspect of a number, such as its weight, cost, temperature, or duration of time

unit price The cost for one measurement of an item, such as the price per pound of meat

utility A business that offers something useful, such as water, electricity, or telephone service to the public

V

vertical Upright; perpendicular to the horizon; for a geometric figure drawn on a page, the vertical dimension is the greatest top-to-bottom measurement

W

wage Payment for work, usually at an hourly rate

withdraw To remove money from an account

withhold To subtract from a gross amount, such as a payment for taxes that is taken from a gross wage

INDEX